METHODS AND PROTOCOLS IN FOOD SCIENCE

Series Editor
Anderson S. Sant'Ana
University of Campinas
Campinas, Brazil

For further volumes:
http://www.springer.com/series/16556

Methods and Protocols in Food Science series is devoted to the publication of research protocols and methodologies in all fields of food science.

Volumes and chapters will be organized by field and presented in such way that the readers will be able to reproduce the experiments in a step-by-step style. Each protocol will be characterized by a brief introductory section, followed by a short aims section, in which the precise purpose of the protocol will be clarified.

Bioactives Encapsulation

Food Applications

Edited by

Samantha Cristina de Pinho

School of Animal Science and Food Engineering (FZEA), University of Sao Paulo (USP), Pirassununga, SP, Brazil

Carmen Silvia Favaro-Trindade

School of Animal Science and Food Engineering (FZEA), University of Sao Paulo (USP), Pirassununga, SP, Brazil

Humana Press

Editors
Samantha Cristina de Pinho
School of Animal Science and Food Engineering (FZEA)
University of Sao Paulo (USP)
Pirassununga, SP, Brazil

Carmen Silvia Favaro-Trindade
School of Animal Science and Food Engineering (FZEA)
University of Sao Paulo (USP)
Pirassununga, SP, Brazil

ISSN 2662-950X ISSN 2662-9518 (electronic)
Methods and Protocols in Food Science
ISBN 978-1-0716-4540-6 ISBN 978-1-0716-4538-3 (eBook)
https://doi.org/10.1007/978-1-0716-4538-3

This Humana imprint is published by the registered company Springer Science+Business Media, LLC, part of Springer Nature.
The registered company address is: 1 New York Plaza, New York, NY 10004, U.S.A.

Preface to the Series

Methods and Protocols in Food Science series is devoted to the publication of research protocols and methodologies in all fields of food science. The series is unique as it includes protocols developed, validated, and used by food and related scientists. It also provides theoretical basis for each protocol. Aspects related to improvements in the protocols, adaptations, and further developments in the protocols may also be approached.

Methods and Protocols in Food Science series aims to bring the most recent developments in research protocols in the field as well as very well established methods. As such the series targets undergraduates, graduates, and researchers in the field of food science and correlated areas. The protocols documented in the series will be highly useful for scientific inquiries in the field of food sciences, presented in such way that the readers will be able to reproduce the experiments in a step-by-step style.

Each protocol will be characterized by a brief introductory section followed by a short aims section, in which the precise purpose of the protocol is clarified. Then, an in-depth list of materials and reagents required for employing the protocol is presented, followed by a comprehensive and step-by-step procedure on how to perform that experiment. The next section brings the dos and don'ts when carrying out the protocol, followed by the main pitfalls faced and how to troubleshoot them. Finally, template results will be presented and their meaning/conclusions addressed.

The Methods and Protocols in Food Science series will fill an important gap, addressing a common complain of food scientists, regarding the difficulties in repeating experiments detailed in scientific papers. With this, the series has a potential to become a reference material in food science laboratories of research centers and universities throughout the world.

Campinas, Brazil *Anderson S. Sant'Ana*

Preface

This book intends to be a contribution to the field of microencapsulation applied to food development. The various microencapsulation techniques have been fundamental for protection of several types of bioactive molecules of high importance in food, as vitamins, pigments, antioxidants, antimicrobials, and flavorings. Such protection can be useful to provide food products of higher nutritional value or to allow the use of the encapsulated ingredients in their preservation. Microencapsulation techniques are supposed to be capable to create conditions for controlled release, alteration of solubility, enhancement of bioaccessibility and bioavailability, as well as protection of the bioactives from possible deleterious interactions with other components present in the food formulation and from processing conditions (Calderon-Oliver & Ponce-Alquicira, 2022). Other possible reasons for microencapsulation are more practical, such as facilitating handling or dosing in small quantities. Microencapsulating bioactive compounds is capable to benefit the food and beverage industries as more precise and efficient encapsulation technologies constantly emerge from scientific studies.

Regarding functional foods, market reports indicate that consumers are more open to consume this type of product. Such market is expected to reach more than 390 billion dollars by 2027. As for them market for nutraceuticals, their global market size is projected to reach USD 136.1 billion by 2029, displaying an annual rate of 5.3% in the period 2024–2029.

Even the use of Artificial Intelligence is a factor in the fast expansion of this market, as it has been used to optimize the discovering and production of bioactives. The use of AI to help unravel the gut microbiome interactions has led to faster development of probiotics, as well as prebiotics and symbiotics bioactives.

Therefore, in order to be in the same pace of the market needs, the constant technological development of microencapsulation techniques is required. Traditionally, this is an area of study in which the scientific production is quite high and diverse, reflecting the high dynamics of a continuous need of new food products' development and progress in the understanding of how bioactives act and can enhance the human health.

The definition of microencapsulation is it is the process for coating bioactive ingredients in the form of small structures, with dimensions ranging from nanometers to micrometers. These structures can be composed of homogeneous or heterogeneous matrices, which are small capsules that allow the protection of the active ingredients inside them. The microencapsulation techniques encompass a set of steps or conditions required to protect the bioactive ingredients of solid, liquid, and gaseous materials in micro- or nano-capsules. These capsules also have the ability of releasing their contents under conditions and quantities through the time, leading to controlled release, alterations in the solubility, and enhancement of the bioaccesibility and bioavailability of the encapsulated compounds.

The cost is one of the most important factors that will determine the applicability of the microencapsulation process for a certain bioactive and product. Lower microcapsule production costs will allow for maintaining a competitive price without compromising product quality. Three aspects must be considered in the industrial and commercial levels for microencapsulation: (i) the production cost of the microcapsules; (ii) the possibility of microcapsules to be incorporated in different food products; (iii) the possibility of scaling up from proof of concept to large-scale production. As for the third aspect, this book can be used as a reference in the development of lab-scale microencapsulation processes, as the level of detail in which the techniques are described can be useful in making such development faster.

Pirassununga, Brazil *Samantha Cristina de Pinho*
São Paulo, Brazil *Carmen Silvia Favaro-Trindade*

Contents

Contributors

IZABELA DUTRA ALVIM • *Center of Cereal and Chocolate Technology, Institute of Food Technology (ITAL), Campinas, SP, Brazil*

MIGUEL A. CERQUEIRA • *International Iberian Nanotechnology Laboratory (INL), Braga, Portugal*

MATHEUS ANDRADE CHAVES • *Department of Veterinary Medicine (ZMV), School of Animal Science and Food Engineering (FZEA) - University of São Paulo (USP), Pirassununga, SP, Brazil*

TALITA ALINE COMUNIAN • *Teagasc Food Research Centre, Moorepark, Fermoy, Co., Cork, Ireland*

ROSIANE LOPES DA CUNHA • *Department of Food Engineering and Technology, School of Food Engineering (FEA) - State University of Campinas (UNICAMP), Campinas, São Paulo, Brazil*

ÁLVARO GUSTAVO FERREIRA DA SILVA • *Department of Food Science and Nutrition, School of Food Engineering, State University of Campinas (UNICAMP), Campinas, SP, Brazil*

MÁRCIA REGINA DA SILVA PEDRINI • *Department of Chemical Engineering, Federal University of Rio Grande do Norte (UFRN), Natal, RN, Brazil*

MARLUCI PALAZZOLLI DA SILVA-PADILHA • *Department of Food Engineering and Technology, School of Food Engineering (FEA) - State University of Campinas (UNICAMP), Campinas, SP, Brazil*

EDUARDO WAGNER VASCONCELOS DE ANDRADE • *Department of Chemical Engineering, Federal University of Rio Grande do Norte (UFRN), Natal, RN, Brazil*

FÁBIO GONÇALVES MACÊDO DE MEDEIROS • *Department of Chemical Engineering, Federal University of Rio Grande do Norte (UFRN), Natal, RN, Brazil; Plants for Human Health Institute, Department of Food, Bioprocessing and Nutrition Sciences, North Carolina State University (NCSU), Kannapolis, NC, USA*

SILVIA CRISTINA SOBOTTKA ROLIM DE MOURA • *Department of Chemical Engineering, Federal University of Rio Grande do Norte (UFRN), Natal, RN, Brazil*

VOLNEI BRITO DE SOUZA • *Center of Biological and Health Sciences, Federal University of Oeste da Bahia (UFOB), Barreiras, BA, Brazil*

CYNTHIA DITCHFIELD • *School of Animal Science and Food Engineering (FZEA) - University of Sao Paulo (USP), Pirassununga, SP, Brazil*

JESSICA THAÍS DO PRADO SILVA • *São Paulo State University (UNESP), Institute of Biosciences, Humanities and Exact Sciences, São José do Rio Preto, SP, Brazil*

CARMEN SILVIA FAVARO-TRINDADE • *School of Animal Science and Food Engineering (FZEA), University of Sao Paulo (USP), Pirassununga, SP, Brazil*

CRISTHIAN RAFAEL LOPES FRANCISCO • *Department of Food Engineering and Technology, School of Food Engineering - State University of Campinas (UNICAMP), Campinas, SP, Brazil*

CHARLES WINDSON ISIDORO HAMINIUK • *Federal Technological University of Paraná (UTFPR), Curitiba, PR, Brazil*

ROBERTA TARGINO HOSKIN • *Department of Chemical Engineering, Federal University of Rio Grande do Norte (UFRN), Natal, RN, Brazil; Plants for Human Health Institute,*

Department of Food, Bioprocessing and Nutrition Sciences, North Carolina State University, Kannapolis, NC, USA

JÚLIA CRISTINA JOSÉ • *School of Animal Science and Food Engineering (FZEA), University of Sao Paulo (USP), Pirassununga, SP, Brazil*

GISELLE MARIA MACIEL • *Federal Technological University of Paraná (UTFPR), Curitiba, PR, Brazil*

VASCO D. F. MARTINS • *International Iberian Nanotechnology Laboratory (INL), Braga, Portugal*

MARIANO MICHELON • *School of Chemistry and Food Engineering, Federal University of Rio Grande (FURG), Rio Grande, RS, Brazil*

IZABEL CRISTINA FREITAS MORAES • *School of Animal Science and Food Engineering (FZEA), University of Sao Paulo (USP), Pirassununga, SP, Brazil*

VÂNIA REGINA NICOLETTI • *São Paulo State University (UNESP), Institute of Biosciences, Humanities and Exact Sciences, São José do Rio Preto, São Paulo, Brazil*

LORENZO M. PASTRANA • *International Iberian Nanotechnology Laboratory (INL), Braga, Portugal*

SAMANTHA CRISTINA DE PINHO • *School of Animal Science and Food Engineering (FZEA), University of Sao Paulo (USP), Pirassununga, SP, Brazil*

TIAGO CARREGARI POLACHINI • *São Paulo State University (UNESP), Institute of Biosciences, Humanities and Exact Sciences, São José do Rio Preto, SP, Brazil*

SANTIAGO RAMÍREZ-LÓPEZ • *Faculty of Engineering and Architecture, Peruvian Union University, Lima, Peru*

GISELLE VALLIM CORRÊA RAMOS • *Faculty of Animal Science and Food Engineering, Department of Food Engineering (ZEA), School of Animal Science and Food Engineering, University of São Paulo (USP), Pirassununga, SP, Brazil*

ELISA FRANCO RIBEIRO • *São Paulo State University (UNESP), Institute of Biosciences, Humanities and Exact Sciences, São José do Rio Preto, SP, Brazil*

MARTIN DANIEL AYA RODRIGUEZ • *Department of Food Science and Nutrition, School of Food Engineering (FEA) - State University of Campinas (UNICAMP), Campinas, SP, Brazil*

FERNANDA THAÍS VIEIRA RUBIO • *Department of Chemical Engineering , Polytechnical School (EP) - University of Sao Paulo (USP), São Paulo, SP, Brazil*

PRISCILLA DAYANE DE FREITAS SANTOS • *Department of Food Engineering and Technology - School of Food Engineering (FEA), State University of Campinas (UNICAMP), Campinas, SP, Brazil*

TATIANA PORTO SANTOS • *Laboratory of Food Process Engineering, Wageningen University and Research, Wageningen, WG, The Netherlands*

SAMUEL HENRIQUE GOMES DE SÁ • *School of Animal Science and Food Engineering (FZEA), University of Sao Paulo (USP), Pirassununga, SP, Brazil*

PEDRO M. SILVA • *International Iberian Nanotechnology Laboratory (INL), Braga, Portugal*

GUILHERME MIRANDA TAVARES • *Department of Food Science and Nutrition, School of Food Engineering (FEA), State University of Campinas (UNICAMP), Campinas, SP, Brazil*

FABRICIO LUIZ TULINI • *Center of Biological and Health Sciences, Federal University of Oeste da Bahia (UFOB), Barreiras, BA, Brazil*

Chapter 1

Complex Coacervation for Encapsulation of Bioactive Compounds, Nutrients, and Probiotics

Marluci Palazzolli da Silva-Padilha, Talita Aline Comunian, and Carmen Silvia Favaro-Trindade

Abstract

Coacervation is one of the earliest microencapsulation techniques, which has been commercially used since the 1950s. Complex coacervation involves the interaction of oppositely charged polymers in aqueous form. Gelatin-Arabic gum coacervates have been successfully produced and used in many applications, especially as vehicles for lipophilic molecules. Recently, successful strategies to encapsulate hydrophilic molecules by complex coacervation were investigated, including the production of double emulsions as a previous step before the coacervation. This technique also encapsulated particles and probiotics based on their previous immobilization in high melting point vegetable fats. Concerning polymeric materials, novel proposals based on plant proteins are growing due to flexitarianism. Also, food by-products are sources of polysaccharides, such as pectins and gums, which are sustainable polyelectrolytes to produce complexes. Complementary studies of release mechanisms are relevant to avoid the premature expulsion of bioactive compounds after incorporation in food matrixes and in vitro digestion. The scalability of complex coacervation in the food industry still needs to be explored, but combining a physical method, such as spray-nozzle, could enhance their feasibility. Thus, it is unquestionable that complex coacervation is a multifaceted microencapsulation technique, and the advances in colloid technology reflect their potentiality for the protection of bioactive compounds for food incorporation, besides many other functionalities. This protocol intends to guide the reader along the key steps to complex coacervation encapsulation, including the selection of polymers, operational conditions, the techniques for coacervate evaluation, as well as give an overview of the most critical influencing factors (e.g., concentration and proportion of biopolymers, pH, the concentration of bioactive compound, homogenization conditions, and temperature). It provides tips on how to deal with some troubleshooting, such as culturing probiotic cells before the microencapsulation, procedures for disrupting coacervates, and poor stability over shelf-life or under environmental/processing conditions.

Key words Microencapsulation, Colloid, Lipophilic, Hydrophilic, Technological modification, Agri-food waste

1 Introduction

Microencapsulation techniques involve the production of tiny "packages," called microcapsules, microspheres, or microparticles, which consist of structures that have a core covered or immobilized

Samantha Cristina de Pinho and Carmen S.F. Trindade (eds.), *Bioactives Encapsulation: Food Applications*, Methods and Protocols in Food Science, https://doi.org/10.1007/978-1-0716-4538-3_1,

by one or more polymers [1]. These structures have numerous functionalities and have been employed for decades to solve or minimize difficulties faced by food producers, such as the high volatility and chemical instability of aromas, the high reactivity of some vitamins, the low water-solubility and the tendency to rapidly oxidize omega-3 fatty acids, unpleasant taste of vegetable extracts, and oxygen susceptibility of probiotics.

Many techniques can be applied to encapsulate food nutrients, additives, and bioactive compounds, such as spray-drying, cooling/chilling/congealing, ionic gelation, and liposomes. In this chapter, emphasis will be given to the complex coacervation technique, first described in 1929 by Bungenberg de Jong and Kruyt. It has been used since 1954 when the National Cash Register Company began commercializing carbonless copy paper. Since then, coacervation has been intensively studied and commercially used in different areas.

The technology of complex coacervation can be described as a phase separation phenomenon induced by electrostatic interactions between polymers of opposite electrical charges. In this context, there are many pairs of biopolymers suitable for electrostatic interactions; however, gelatin—an amphoteric biopolymer carrying by carboxyl and amino groups—and Arabic gum—an always negatively charged biopolymer—form the classic, so the most used pair for complex coacervation. Based on this, coacervates can be described as electrostatic complexes; however, hydrogen bonding and hydrophobic interactions can also occur during coacervation.

Coacervates or microcapsules produced by complex coacervation may have single or multiple cores. Single-core coacervates generally have an oval shape, while coacervates with multiple cores can also have an oval or an indefinite shape, depending on many factors, especially the polymers used for complex production.

Although complex coacervation can occur in aqueous or organic liquids, it is promoted mainly in aqueous mediums for food applications. It is also used mainly to encapsulate hydrophobic compounds by hydrophilic-charged biopolymers. However, hydrophilic molecules (such as casein peptides, ascorbic acid, sucralose, and lactase) were successfully encapsulated using a double emulsion (W/O/W) before complex coacervation steps [2–4]. Recently, another approach involves the direct incorporation of cinnamon dry extract into biopolymers, adjusting the pH to promote electrostatic interactions [5].

This technique was also used to encapsulate two compounds (xylitol and menthol) simultaneously into just one structure or coacervate, improving the functionality of individual compounds and maximizing their performance [6] and also to encapsulate simultaneously probiotics and proanthocyanidin-rich cinnamon extract [7], which can be released together into the gut, presenting the possibility of acting in synergy. Other innovations were the encapsulation of liquid propolis extract [8] and solid particles of cinnamon extract using gelatin and different polysaccharides [5, 9] and probiotics by their immobilization into solid lipid particles followed by these particles covered by protein–polysaccharide complex [7, 10].

Indeed, many bioactive molecules and probiotics would benefit from being encapsulated, and this possibility has increased the market and consumption of functional foods. Some interesting studies had promised results using complex coacervation to encapsulate bioactive compounds in this context. For instance, Gomez-Estaca et al. [11] encapsulated astaxanthin-containing lipid extract from shrimp using a novel gelatin–cashew gum complex and applied it to yogurt. The yogurts were evaluated by a trained sensory panel, which confirmed the improvement in coloring capacity achieved through encapsulation, although no differences in odor were found. Similarly, Comunian et al. [12] also produced yogurt with coacervates containing echium oil (a source of omega-3 fatty acids), phytosterol, and sinapic acid produced by complex coacervation using the pairs gelatin–Arabic gum and gelatin–cashew gum. The microcapsules did not compromise the product's rheological properties and physicochemical stability, and the oil became more stable in the products when encapsulated than free.

Complex coacervation offers several advantages for encapsulating bioactive compounds. These include versatility, relatively low cost, simplicity, and the use of mild operational conditions, such as low temperatures and water as a solvent. Additionally, it enables the formation of reservoir-type or true microcapsule structures with double walls, which can release their content in a controlled manner at specific sites. Complex coacervation also provides high encapsulation efficiency, particularly for lipophilic compounds. However, it also has some drawbacks since it is a batch process with low productivity, and, in general, it is necessary to conjugate a crosslinker (for instance, transglutaminase) for the production of more stable coacervates and a dehydration process to produce microbiological stable powders.

Thus, this chapter aims to introduce the basis of this method of encapsulation and guide the reader through the main steps for the production of coacervates, which can be vehicles for compounds with different properties.

2 Materials

Gelatin and gum Arabic is the classical polymer pair for microencapsulation by complex coacervation. Considering the growing demand for vegan ingredients, investigating suitable biopolymers for producing coacervates is a promising field to explore emerging gums and proteins. Table 1 lists the most used proteins and carbohydrate combinations for complex coacervation.

Table 1
Examples of some combinations of wall materials for complex coacervation

Combination of wall materials	References
Gelatin and sodium hexametaphosphate	[13]
Whey protein and gum Arabic	[14]
Gelatin and gum Arabic; gelatin and cashew gum	[15]
Gelatin and gum Arabic	[3, 16]
Gelatin with gum Arabic, pectin, cashew tree gum, carboxymethylcellulose, or κ-carrageenan	[9]
Soluble pea protein and gum Arabic	[17]
Gelatin and carrageenan	[18]
Lactoferrin and carboxymethyl tara gum	[19]
Beta-lactoglobulin and gelatin	[20]
Soy protein isolate, sodium caseinate, sodium carboxymethylcellulose, sodium alginate	[21]
Bovine serum albumin and gum acacia	[22]
Gelatin and pectin	[23]
Isolated soy protein and pectin	[8]
Casein and pectin	[24]

However, before using these proteins and/or carbohydrate combinations to encapsulate bioactive compounds, it is essential to evaluate their physicochemical properties to achieve the highest possible encapsulation efficiency and yield. As a result, the following sections provide extensive characterization assessments of raw materials (wall materials).

3 Methods

3.1 Characterization of Raw Materials Before Their Use for Coacervation

(i) **Solubility Curve for Proteins**

Protein is one of the most commonly utilized materials for complex coacervation, so understanding its behavior and solubility is critical. Proteins' solubility varies with pH, and their curve is U-shaped [17, 25, 26]. In other words, protein solubility decreases as it approaches its isoelectric point and increases below or above it, with maximal solubility occurring at high pH values.

To generate the protein solubility curve, aliquots of protein solutions (in MilliQ water) with the same concentration are prepared at various pH levels, typically ranging from 2 to 8. Increasing the pH values by 0.5 is necessary for a more accurate result. To ensure that the protein is thoroughly hydrated, the solutions should be prepared one day before and stored at 4 °C overnight. The protein in these solutions is quantified, and the results are called total protein. The aliquots are centrifuged at 10,000×*g* for 20 min at room temperature. Each solution's supernatant is quantified again based on the protein level, and the results are referred to as the soluble protein. The solubility of proteins is estimated using Eq. 1.

$$\text{Solubility}\ (\%) = (\text{Soluble Protein} \div \text{Total Protein}) \times 100. \qquad (1)$$

(ii) **Zeta Potential Measurements**

Complex coacervation is an encapsulating technology that relies on the electrostatic interactions between two materials. As a result, understanding the charge of the materials is critical for determining the optimal conditions for high-yield coacervate formation. Zeta potential measurement is the most known and straightforward way to investigate charges [27, 28]. In the case of a protein–polysaccharide combination, the protein typically has a positive charge and the polysaccharide a negative charge at specific conditions. Similarly, as mentioned in obtaining the protein solubility curve, the charge of wall materials is evaluated for their solutions at various pHs (from pH 2.0 to pH 8.0). The results of charge versus pH values are plotted and compared. The most significant yield of coacervation production is expected to occur at pH values where the charges of both components are neutralized when combined.

(iii) **Turbidity of the Wall Materials Mixture**

Turbidity measurement is one of the most commonly used methods for quantifying coacervate production. Solutions containing wall materials are prepared at low concentrations (0.1% w/v), and various protein:polysaccharide ratios are investigated (e.g., 1:1, 2:1, 3:1, 4:1, 10:1 w/w and/or vice versa). These ratios vary depending on the materials utilized. Preliminary testing should determine the suitability of each combination of wall materials. Each combination is constantly stirred, and the pH is adjusted by adding HCl or NaOH. A UV/vis spectrophotometer is used to measure various ratios and pH values at room temperature at 600 nm, and the turbidity is calculated using Eq. 2 [27–29].

$$\text{Turbidity} = -\ln\ (T), \qquad (2)$$

where T is the transmittance.

The maximal formation of coacervates corresponds to the highest turbidity levels. In other words, when no complexes or aggregates are formed, there is no absorption at 600 nm, and low turbidity levels are observed.

Another correlation can be obtained between turbidity and pH levels. It is known that the formation of coacervates consists of three steps: (1) the beginning of the production of insoluble coacervates, (2) when the formation of coacervates reaches its highest level, and (3) when the formation of insoluble coacervates decreases and soluble ones form. In this scenario, the pH values for each phase are identified as pH_1, pH_{opt}, and pH_2, respectively. Various authors mention this concept [17, 27, 30, 31].

(iv) **UV/Vis Absorbance of the Wall Materials Mixture**

As mentioned, the most used combination of wall materials are proteins and polysaccharides. Proteins are composed of amino acids (AAs), and the aromatic AAs (tryptophan, tyrosine, and phenylalanine) suffer modification when applied to different pH values and temperatures. This modification can be investigated and understood by measuring the changes in absorbance [27]. The formation of coacervates evolves changes in the pH and sometimes the application of heating. Hence, the interaction of proteins and polysaccharides during the complex coacervation process can be studied by measuring the absorbance of their mixtures. In this case, solutions combining proteins and polysaccharides are prepared with the same protein concentration but with varying concentrations of polysaccharides. The absorbance is measured at 280 nm using a UV/vis spectrophotometer.

The output expected from this measurement is higher absorbance values for the higher polysaccharide concentrations. According to Chen et al. [32], it happens due to aromatic amino acid residue exposure. In addition, different polysaccharides may be compared to investigate the combination with the highest yield of coacervation with the protein since the combination with the highest absorbance values presents the strongest interactions.

(v) **Thermodynamics and Molecular Interactions Between Wall Materials**

The molecular interactions between proteins and polysaccharides are studied according to their thermodynamic parameters, which can be obtained by different methods. Isothermal titration calorimetry and circular dichroism are two sensitive methods that have been applied to monitor the molecular interactions and changes in the secondary structure of proteins when used as wall material for complex coacervation [26, 28, 33, 34].

However, the measurement of fluorescence intensities of the combination protein:polysaccharides at different concentrations using a fluorescence spectrophotometer can also be a simple alternative to calculate the thermodynamic parameters [17, 35, 36]. As mentioned, for the UV/vis absorbance of the wall materials mixture, solutions with the same protein content but different polysaccharide concentrations are prepared. The fluorescence intensities are measured at excitation wavelength at 280 nm and emission spectra at 290–500 nm. The measurements are done at three different temperatures, usually 288, 298, and 310 K, but other temperatures could also be analyzed.

According to Perez et al. [36], Ying et al. [37], and Vijayan et al. [35], the first step for the calculation of the thermodynamic parameters is to obtain the dynamic quenching constant according to Eq. 3:

$$Fo \div F = 1 + Ksv.[Q], \tag{3}$$

where

Fo = fluorescence intensity of the protein solution alone

F = fluorescence intensity of the protein–polysaccharide solutions

$[Q]$ = concentration of the polysaccharide (also known as quencher) in mol/L

After that, the binding constant (Ka) of the combination of protein–polysaccharide and the number of binding sites per protein (n), indicative of the intensity and the number of protein molecules bound to the polysaccharide, is obtained according to Eq. 4:

$$\log \frac{Fo - F}{F} = \mathrm{logKa} + n\log[Q]. \tag{4}$$

Finally, the Gibbs free energy change (ΔG) and the entropy change (ΔH) are obtained according to Eqs. 5 and 6, and the enthalpy and entropy changes (ΔS) are obtained from the curve of lnKa versus $1/T$:

$$\log\ \mathrm{Ka} = -\Delta H/2.303\mathrm{RT} + \Delta S/2.302\mathrm{R} \tag{5}$$

$$\Delta G = \Delta H - T\Delta S = -\mathrm{RTlinKa}, \tag{6}$$

where

T = temperature

R = ideal gas constant (8.314 $\mathrm{J.K^{-1}.mol^{-1}}$)

Due to the interaction of the polysaccharides (quencher) with the protein by either collisional quenching or coacervate formation, it is expected that the fluorescence intensity of the mixtures decreases with the increase of $[Q]$ [17]. Moreover, with the values of the free Gibbs energy, enthalpy, and entropy changes, it is possible to understand the main driving forces of the interactions between the protein and polysaccharides, such as electrostatic interactions, hydrogen bonds, and van der Waals forces [38].

3.2 Complex Coacervation

Complex coacervation is an encapsulation method based on the ability of cationic and anionic water-soluble polymers to interact electrostatically, resulting in a polymer-rich phase called coacervate [39]. A typical microencapsulation process of a hydrophobic core (e.g., oils, carotenoids, and aromas) based on complex coacervation of Arabic gum and gelatin is shown in Fig. 1.

For the encapsulation of water-immiscible cores, basic complex coacervation steps are (i) production of an O/W emulsion using gelatin (an amphoteric substance) as the surfactant; gelatin dispersion pH value should be above its isoelectric point, so carboxyl groups will be ionized, giving this protein a negative charge; (ii) gently to blend the emulsion with a gum Arabic (an anionic polysaccharide) dispersion; (iii) pH reduction, aiming to change gelatin charges by neutralizing its carboxyl groups and forming positively charged amino groups; (iv) water dilution of the mixture; (v) coacervation, mainly by the electrostatic interaction of positively charged amino groups of gelatin and negatively charged anionic polysaccharide groups; in addition, hydrogen bonds and hydrophobic interactions also can occur; (vi) cooling of the systems; (vii) cross-linking, generally using transglutaminase; and (viii) coacervates separation by decantation, filtration, or centrifugation. In addition, many coacervate systems are also dehydrated by spray- or freeze-drying.

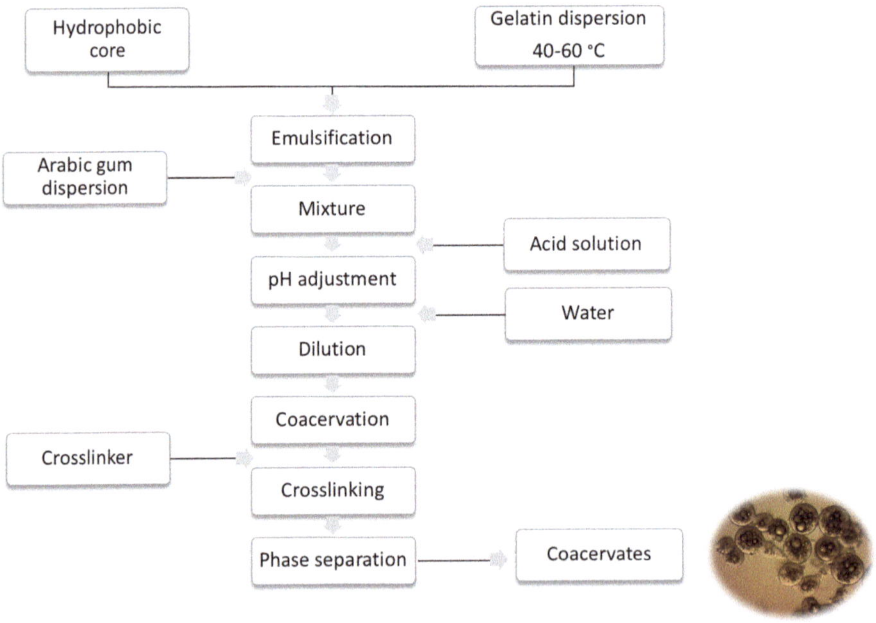

Fig. 1 Rationale for the microencapsulation process of a hydrophobic core by complex coacervation of Arabic gum and gelatin

As previously stated, complex coacervation can encapsulate various materials, including lipophilic and hydrophilic bioactive compounds, probiotics, and solid lipid particles. If the compound to be encapsulated is hydrophilic, a water-in-oil emulsion is formed. In this case, the compound is added into the inner aqueous phase. The first biopolymer, used as a wall material in an aqueous solution, is combined with the simple emulsion to form a double water-in-oil-in-water emulsion. As mentioned, complex coacervation occurs when the second biopolymer is added, and the pH is adjusted. The steps for encapsulating these materials are provided below.

3.2.1 Preparation of Bioactive Molecules for Microencapsulation

Generally, the solubility of bioactive molecules in water and organic solvents is well known and described in the product specification data sheet. In addition, the physicochemical properties of molecules, such as the n-octanol-water partition coefficient (log P), estimated the hydrophobicity of bioactive molecules.

Hydrophobic bioactive compounds can be either directly incorporated into the lipid matrix, which consists of oil or solid fat, or emulsified in a protein dispersion before microencapsulation via complex coacervation. The time interval between the preparation of bioactive molecules and encapsulation should not be long, considering the possibility of bioactive molecules degrading when exposed to light and oxygen.

Concerning the hydrophilic molecules preparation, there are three strategies used for microencapsulation by complex coacervation: (i) double emulsion by two-step emulsification, (ii) direct incorporation in protein solution, and (iii) immobilization of hydrophilic molecules into solid lipid matrix. The following topics indicate the procedures for incorporating lipophilic and hydrophilic bioactive compounds:

(i) **Hydrophobic Bioactive Molecules: Direct Incorporation**

- Stock the hydrophobic samples protected from heat and light, such as fish and vegetable oils and other bioactive molecules.
- Incorporate the hydrophobic molecules, such as liposoluble vitamins and phenolics, in the oil by magnetic stirring for at least 30 min.
- After homogenization, stock the oil loaded with bioactive compounds until microencapsulation is used.

(ii) **Hydrophilic Bioactive Molecules**

Procedure 1: Double emulsion by two-step emulsification to incorporate hydrophilic molecules [3, 40, 41]

- Prepare the primary (W/O) emulsion using adequate concentration of hydrophilic bioactive molecules in distilled water, vegetable oil, and emulsifiers. Preliminary tests are required to select the best ratios of each component and processing parameters (speed/time) to produce a stable primary emulsion.

- Homogenize the mixture of bioactive compounds, distilled water, oil, and emulsifiers using an Ultra-Turrax at 12,000 rpm for 4 min.
- Verify the stability of the primary emulsion over time, considering the procedure intervals until microencapsulation.

Procedure 2: Direct incorporation of hydrosoluble molecules powder [42]

- Add the hydrosoluble molecules in powder form directly into distilled water, using the proportion twice of encapsulating solution volumes. It is recommended that some trial tests be performed to select an adequate concentration.
- Homogenize the mixture using an Ultra-Turrax at 6000 rpm for 2 min.
- Proceed to incorporate the protein solution for microencapsulation by complex coacervation.

Procedure 3: Immobilization of bioactive compounds into melted fat [7, 10]

This procedure could be used for hydrophobic and hydrophilic cells and compounds because it is an immobilization to produce solid lipid particles. The sequential step is coating lipid particles using the electrostatic interaction of polysaccharides and proteins. In this way, the immobilization procedure is described below:

- Heat the fat until completely melting, and keep it under heat for 30 min.
- Incorporate probiotic inoculum or bioactive compounds into melted fat, followed by homogenization using an Ultra-Turrax at 3500 rpm for 60 s.
- Use a water bath to control the temperature for melting fat and homogenization.

Proceed to incorporate the protein solution, maintaining the heat.

3.2.2 Preparation of Probiotic Inoculum

Probiotic cells' hydrophilic or hydrophobic profile varies depending on the strain, membrane cell composition, growth conditions, etc. Since the complex coacervation technique is efficient for protecting lipophilic compounds, determining the hydrophobicity of cells is a crucial step to devise strategies for protecting probiotics. Currently, two approaches are widely used in the literature: (1) producing a primary emulsion and (2) direct incorporation into the lipid matrix.

(i) **Culturing of Probiotic Cells**

- Prepare microbiological broth and agar, referring to the probiotic strain used in the desired study, using a sterilizer at 121 °C at >15 psi for 15 min.
- Decontaminate the work environment using UV light in a biosafety cabinet.
- Prepare probiotic inoculum following suppliers' recommendations, considering the viability of freeze-dried cells.
- Culturing probiotic cells [16]: Add a small amount of freeze-dried probiotic cells, around 50 mg, to 10 mL of De Man, Rogosa, and Sharpe—MRS broth for incubation at 37 °C for 18 h. Add the inoculum to 100 mL MRS broth and incubate at 37 °C for 18 h. Collect the cells by centrifugation (6603×*g*, for 5 min at 4 °C), followed by washing twice with 2% sodium citrate or phosphate-buffered saline—PBS buffer.
- *Troubleshooting:* Anaerobic probiotic strains, such as Bifidobacterium spp., require incubation in anaerobic jars and sachets, conditioning an anaerobic system for multiplication.

(ii) **Hydrophobicity Profile of Probiotic Cells**

- Wash the cell pellet twice with potassium hydroxide (KOH, 50 mM, pH 7.0), followed by the suspension with potassium nitrate (KNO_3, 0.1 mM, pH 6.2).
- The suspension (approximately 4 mL) is placed in contact with 1 mL of each solvent (chloroform, xylene, and ethyl acetate) for 5 min.
- Homogenize the suspension for 2 min using a vortex.
- Rest the tubes for 1 h, promoting the phase separation.
- Measure the absorbance of the aqueous phase at 600 nm.
- Calculate the affinity percentage of probiotic cells in the solvents using Eq. 7.

$$\text{Hydrophobicity } (\%) = \left(\frac{OD_A - OD_B}{OD_B}\right) \times 100 \tag{7}$$

In this equation, OD_A is the optical density of the cell suspension in KNO_3 using the KNO_3 as solution blank. OD_B is the optical density of the cell suspension in KNO_3 and solvent, using the KNO_3 solution as blank with solvent.

- *Troubleshooting:* Prepare a cell suspension solution with absorbance around 0.8 to ensure an adequate concentration of cells to adhere to the solvents and, at the same time, to avoid precipitation.

The results suggest that hydrophilic probiotic cells may be incorporated into a previous emulsification step or immobilized into melted solid fat. In contrast, hydrophobic cells may be directly added to the lipid matrix.

3.2.3 Procedures for Microencapsulation of Probiotics and Bioactive Compounds by Complex Coacervation

Several protocols have been developed for microencapsulation by complex coacervation. Over the years, some preliminary modifications to the process have been proposed, such as the production of double emulsification [3, 40] and the immobilization of bioactives into a melted lipid matrix [10]. An additional drying step is required when the coacervates need to be stored for a long period or will be applied to food matrices with low water activity. In this way, freeze-drying and spray-drying are both approaches that are used the most.

The protocol presented below is proposed for immobilizing bioactive compounds with varying polarity in vegetable fat, followed by covering with biopolymers through complex coacervation [10].

- *Preparation of encapsulating materials:* Prepare 150 mL of gum Arabic solution (2.5% w/v) using ultrapure water and a magnetic stirrer at 200 rpm. Prepare 150 mL of porcine gelatin solution (2.5% w/v) using a magnetic stirrer at 200 rpm. Heat the solution at 60 °C to improve the gelatin dissolution. Store the gum Arabic and gelatin solutions at refrigeration temperature for 24 h to complete the hydration process.
- *Incorporation of probiotic cells into melted fat followed by covering with encapsulating materials:* Weigh 5 g of solid fat and place in a small water bath to melt for at least 30 min. Take the cell mass (discard the citrate/PBS buffer) and mix it in the fat. Use the Ultra-Turrax at the lowest speed (3500 rpm) to homogenize the cell mass in fat. Keep homogenizing for as little time as possible. Remove the encapsulating materials from the refrigerator and equilibrate the temperature using a water bath at 60 °C. Add the melted fat and probiotics mixture into the gelatin solution using an Ultra-Turrax at 7000 rpm for 90 s. Take a larger beaker and place the emulsion. Add 150 mL of Arabic gum solution and stir with a magnetic stirrer. Adjust the pH of this mixture to 4.5 with NaOH (1 M) and keep it under magnetic stirring. Add 600 mL of distilled water; this proportion is always twice the encapsulating solution volume. Keep stirring with a magnetic stirrer. Prepare an ice bath using a larger beaker (2.5 L) to place the mixture inside, maintaining the magnetic stirring until it reaches the temperature of 10 °C. Check the formation of complexes using optical microscopy. Remove the beaker from the ice bath and store it in the refrigerator for the decantation of coacervates. Remove the water with a large pipette equipment or with the aid of a hose.

- *Drying of coacervates:* Distribute the resulting coacervates in petri dishes to prepare a thin layer or place the capped tubes containing the coacervates in a horizontal position to increase the surface area, facilitating rapid freezing at low temperatures. Add the frozen samples into the freeze-dryer chamber and turn on the vacuum pump to promote ice removal by sublimation for at least 24 hours. Follow the instructions of the freeze-dryer provider to operate the equipment. Remove the dried samples and transfer them to a desiccator or suitable packaging, protecting them from moisture.

3.2.4 Advantages and Limitations of Complex Coacervation

Like any other encapsulation technique, the complex coacervation offers advantages and limitations. Some examples are shown in Table 2.

3.3 Approaches for Characterization of Coacervates

Many characterization techniques are utilized to study coacervates' development and suitability for food applications. Some examples include microscopy, particle size, encapsulation efficiency and yield, differential scanning calorimetry (DSC), and Fourier transform infrared spectroscopy (FTIR), which are detailed below.

3.3.1 Checking the Microcapsule Morphology by Microscopy and Particle Size Approaches

Microscopy is the most common technique used for studying the morphology of coacervates. Depending on the combination of wall materials, the coacervates may appear spherical or just as an aggregation of polymers. As a result, understanding the interaction of the materials and where the encapsulated component is situated inside these delivery systems is critical to achieving an appropriate food application. In this scenario, various microscopy techniques are

Table 2
Main advantages and limitations of complex coacervation

Advantages	Limitations
It does not use strong conditions, such as high temperature	It is not a continuous process
High encapsulation efficiency for lipophilic compounds	It needs a stable double emulsion for the encapsulation of hydrophilic compounds
Easy and low-cost process	It becomes expensive if freeze-drying is applied
Many biopolymers (proteins and polysaccharides) can be used as wall materials	Extra material (i.e., maltodextrin) is necessary to increase the solid content if spray-drying is applied
Coacervates are stable under gastric conditions, promoting the controlled release of encapsulated compounds	It is not possible to control the particle size. It will depend on the materials and homogenization conditions applied
Without using organic solvents	Limited application into solid food products

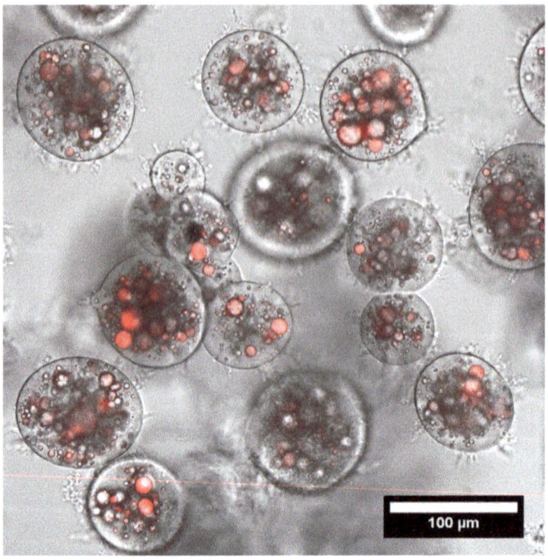

Fig. 2 Confocal microscopy of coacervates formed by combining gelatin and Arabic gum as wall materials with oil as the core

used, including optical, scanning electron microscopy (SEM), and confocal laser scanning microscopy (CLSM). The most basic method is optical microscopy, which requires placing wet samples in a glass slide with an adjusted zoom and lens size. SEM and CLSM, on the other hand, require samples to be prepared beforehand. For SEM, the coacervates must be dried. To differentiate the compounds, CLSM requires the addition of a dye. Lipids and proteins are typically identified using this method [1]. The confocal microscopy can reveal the distribution of bioactive compounds and their interaction with encapsulating materials. Thus, it is a widely used analytical approach in the encapsulation field (Fig. 2).

To understand the stability, mobility, flowability, and possible applications in food products, particle size and particle size distribution must be investigated and combined with microscopy techniques. Consumers can notice particle sizes larger than 100 µm in their mouth, resulting in product rejection. Moreover, small particle size coacervates are more stable and have better encapsulation efficiency. The most used technique to obtain the particle size is integrated-laser light scattering instruments, which measure $D_{4,3}$ and $D_{3,2}$, the average diameter according to volume and surface, respectively. The measurements can be done with liquid and powder samples, depending on the instrument's specifications. Hence, morphology is a crucial feature to consider for complex coacervation.

3.3.2 Strategies for the Evaluation of Encapsulation Efficiency and Yield

To determine how much of the compounds are encapsulated, free on the surface of the coacervates, and lost during the encapsulation process, encapsulation efficiency and yield are estimated using the

Table 3
Examples of how to break the coacervates and release the encapsulated compounds for further quantification

Wall materials	Encapsulated bioactive compound	Description to break the coacervates	References
Gelatin and gum Arabic	Ascorbic acid	Add 0.4 g of the sample into 10 mL of extraction solution containing 5% NaCl and 1% sodium dodecyl sulfate. Then, use a 1% oxalic acid solution (w/w) containing 0.025% 2,6-dichlorophenolindophenol (w/v) to measure the ascorbic acid levels by titration	[3]
Gelatin, two polysaccharides (Arabic gum and cashew gum) and crosslinkers (sinapic acid and transglutaminase)	Echium seed oil and phytosterols	Add 0.1 g of the sample into a solution containing 5 mL methanol, 5 mL ZnSO4 (5% w/w), and 0.8 g KCl and sonicate for 10 min. Collect the supernatant by centrifugation (4000 rpm/25 °C/10 min) and filter for high-performance liquid chromatography (HPLC) analysis	[15]
Whey protein isolate (WPI) and Arabic gum	Lycopene	Add 250 mg of the powder to 6 mL of 10% NaCl:methanol (ratio of 1: 1). Sonicate for 30 min. Add 30 mL of n-hexane:acetone (ratio of 1:1) to the mixture. Centrifuge the samples at 6000×*g* for 10 min. Collect the supernatant to measure the absorbance by spectrophotometer at 503 nm	[43]
Gelatin and gum Arabic	Guaraná peel extract with probiotics	Add 0.1 g of freeze-dried coacervate to 2.5 mL of 2% sodium citrate solution, followed by incubation at 40 °C/5 min. Sequentially, add 5 mL of hexane to this mixture and sonicate for 5 min. Separate the organic phase by centrifugation (2935 g/5 min). Proceed with the extraction three times to remove the yellow color of the coacervates. Collect the organic phase to quantify the total carotenoids by spectrophotometer UV-Vis	[16]
Gelatin and five different polysaccharides (gum Arabic, pectin, cashew tree gum,	Cinnamon extract	Add 0.1 g of each sample into tubes and 2.5 mL of a 0.1 M NaOH solution. Homogenize the	[9]

(continued)

Table 3 (continued)

Wall materials	Encapsulated bioactive compound	Description to break the coacervates	References
carboxymethylcellulose, and κ-carrageenan)		mixture using a vortex until the complex dissolves. Then, add 5 mL of acetone (AP) acidified with 5% acetic acid in the mixture under slow vortexing. Centrifugate the samples (10 °C/ 6603 g/5 min) and transfer the supernatant to a 25 mL flask. Repeat the procedure with the pellet. Combine the supernatants and the flask volume was quenched with acidified acetone. Use this extract to analyze the total content of phenolic compounds and proanthocyanidins	
Pea protein and three different polysaccharides (carrageenan, tragacanth gum, and xanthan gum)	Blackberry juice	Add 20 mg of particles to 10 mL of ethanol. Homogenize the mixture and protect it from the dark at periodic intervals of 15 min for 1 h. Centrifugate the sample to recover the supernatant to quantify the unencapsulated phenolic compounds (4200×*g*, 25 °C, 15 min). Sonicate the remaining fraction of 9 mL for 10 min, followed by homogenization at 12,000 rpm for 5 min. Maintain the mixture under stirring for 24 h in an orbital homogenizer to complete the release of phenolic compounds	[44]

amount of the bioactive compound measured before and after the structure is ruptured. Encapsulation efficiency and yield are calculated according to Eqs. 8 and 9:

$$\text{Encapsulation efficiency }(\%) = \left[\frac{\text{Total} - \text{Free}}{\text{Total}}\right] \times 100 \qquad (8)$$

$$\text{Encapsulation yield }(\%) = \frac{\text{Total quantified after encapsulation}}{\text{Total added in the process}} \qquad (9)$$

where

Total = total amount of quantified encapsulated compound after the coacervates are broken and the material is released

Free = amount of quantified compound when the coacervates are washed but not broken

Total quantified after encapsulation = Total + Free

Total added in the process = total amount of the bioactive compound added at the beginning of the encapsulation

In other words, encapsulation efficiency calculates the amount of encapsulating material when measured inside and outside the coacervates. Encapsulation yield indicates how much of the compound was lost during the encapsulation process by comparing the total amount quantified after encapsulation to the amount initially introduced.

However, the technique for breaking the coacervates differs depending on the wall materials utilized, with some formations more resistant than others. As a result, Table 3 outlines some of the more often-used processes for rupturing coacervates and releasing encapsulated compounds for further measurement.

3.3.3 Additional Characterization of Coacervates by DSC and FTIR

(i) **Thermal Stability Differential Scattering Calorimetry (DSC)**

Investigating the thermal stability of bioactive molecules after microencapsulation by complex coacervation using differential scattering calorimetry (DSC) may provide useful information for food applications. Among the properties and phenomena that DSC analysis provides, the glass transition temperature (T_g) indicates a transition of material from a glass state to an elastic state [45], in addition to identifying structural modification.

The evaluation of the curves obtained via DSC of the encapsulating materials and coacervates will reveal significant differences due to the electrostatic interaction of biopolymers, as evidenced by the presence of new peaks and the absence of others. These electrostatic interactions influence a new cross-linking network structure between the molecules of the encapsulating materials [46]. Other interesting findings have been reported; for instance, coacervates produced using high molecular weight polysaccharides exhibited high thermal stability. However, exposing them to temperatures above T_g may influence the release of lipid cores containing bioactive compounds [21].

DSC analysis can provide information about the crystalline or amorphous state of bioactive compounds in microcapsules. Consequently, a study evaluated the co-encapsulation of xylitol and menthol via double emulsion followed by complex coacervation using gelatin and gum Arabic [6]. Thermal analysis via DSC of the coacervates did not reveal peaks related to xylitol and menthol, indicating that they were in an amorphous form. The freshness provided by xylitol relies on its crystalline form, whereas the sensation of freshness induced by menthol is independent of its physical state. Therefore, the authors suggested alternative strategies for obtaining microencapsulated xylitol in a crystalline state, such as spray-drying, which yields powders with these characteristics [6].

Thus, DSC analysis contributes to the development of controlled-release coacervates by exploring the thermal properties of biopolymers, their electrostatic interaction, and the state of bioactive compounds in microcapsules. Additionally, this tool can be used to evaluate the efficacy of complex coacervation and the stability of bioactive compounds during storage.

(ii) **FTIR: Fourier-Transform Infrared Spectroscopy**

FTIR is a relevant approach for the chemical characterization of coacervates (1) due to the identification of functional groups and chemical interactions between the biopolymers and bioactive compounds, (2) due to the optimization of complex coacervation process by selecting the biopolymers with electrostatic interaction more stable to improve the protection of bioactive compounds, (3) as it is a complementary tool for monitoring the stability of bioactive compounds after microencapsulation. Some challenges related to FTIR equipment's spectral resolution can limit the ability to distinguish slight differences in spectra, particularly when analyzing complex samples. However, strategies to address this issue are well established in the literature, such as evaluating the spectrum of encapsulating materials and minimizing interference.

The confirmation of microencapsulation by complex coacervation is evaluated by FTIR spectra, considering the presence of amide groups resulting from the interaction between carboxyl groups from polysaccharides and amine groups from protein [41]. Applying different encapsulating materials to form coacervates usually revealed specific interactions between the biopolymers. For instance, FTIR spectrums of coacervates produced using pea protein combined with different polysaccharides, such as carrageenan, tragacanth, and xanthan gums, displayed varied interactions of chemical groups [44]. The FTIR spectrum of the coacervates indicated that the association between pea protein and xanthan gum was more stable, as evidenced by the presence of more intense and well-defined peaks associated with specific interactions between the functional groups of the components [44]. Additionally, the stability of bioactive compounds can be evaluated by FTIR analysis. Another study reported that echium oil and the phytosterol mixture remained intact after microencapsulation by complex coacervation, as indicated by typical peaks representing vibrations of specific functional groups [15].

4 Notes

1. Throughout this chapter, we discussed the importance and main approaches used in complex coacervation, considering it as a versatile microencapsulation technology explored for the protection of bioactive compounds, probiotics, and nutrients.
2. The selection of food-grade biopolymers based on their characterization is an essential preliminary step to provide insights into the process conditions that promote electrostatic interaction.
3. Evaluating the hydrophobicity of bioactive compounds and probiotics reveals adequate procedures for microencapsulation by complex coacervation, such as the direct incorporation in oil, a previous step of emulsification or immobilization in the lipid matrix.
4. Encapsulation efficiency and yield are critical parameters for assessing the operational conditions of complex coacervation.
5. The procedures for disrupting coacervates vary depending on the biopolymers and the binder used and are required to quantify bioactive compounds.
6. Regarding the characterization of coacervates, analytical tools such as optical and confocal microscopies elucidate the morphology. At the same time, DSC and FTIR are complementary assays to confirm the electrostatic interaction and stability of bioactive compounds after the microencapsulation.
7. This chapter provides fundamental practical knowledge of complex coacervation for researchers interested in further exploring this field.
8. Taken together, complex coacervation is a potential technology to protect bioactive compounds, presenting countless research opportunities regarding the optimization of formulations and industrial scalability.

Acknowledgments

Carmen S. Favaro-Trindade thanks Conselho Nacional de Desenvolvimento Científico e Tecnológico (CNPq) for the productivity grant (#310686/2022-9). Talita Aline Comunian received funding from the Research Leaders 2025 programme cofounded by Teagasc and the European Union's Horizon 2020 research and innovation programme under the Marie Skłodowska-Curie grant agreement number 754380. Marluci P. da Silva-Padilha thanks the Sao Paulo Research Foundation (FAPESP) for the scholarship awarded (grant #2016/24895-5 and #2023/02631-0).

References

1. Comunian TA, Favaro-Trindade CS (2016) Microencapsulation using biopolymers as an alternative to produce food enhanced with phytosterols and omega-3 fatty acids: a review. Food Hydrocoll 61:442–457
2. Mendanha DV, Ortiz SEM, Favaro-Trindade CS, Mauri A, Monterrey-Quintero ES, Thomazini M (2009) Microencapsulation of casein hydrolysate by complex coacervation with SPI/pectin. Food Res Int 42:1099–1104. https://doi.org/10.1016/j.foodres.2009.05.007
3. Comunian TA, Thomazini M, Alves AJG, de Matos Junior FE, de Carvalho Balieiro JC, Favaro-Trindade CS (2013) Microencapsulation of ascorbic acid by complex coacervation: protection and controlled release. Food Res Int 52(1):373–379
4. Rocha-Selmi GA, Theodoro AC, Thomazini M, Bolini HMA, Favaro-Trindade CS (2013) Double emulsion stage prior to complex coacervation process for microencapsulation of sweetener sucralose. J Food Eng 119(1):28–32. https://doi.org/10.1016/j.jfoodeng.2013.05.002
5. Souza VB, Thomazini M, Barrientos MAE, Nalin CM, Ferro-Furtado R, Genovese MI, Favaro-Trindade CS (2018) Functional properties and encapsulation of a proanthocyanidin-rich cinnamon extract (Cinnamomum zeylanicum) by complex coacervation using gelatin and different polysaccharides. Food Hydrocoll 77:297–306
6. Santos MG, Carpinteiro DA, Thomazini M, Rocha-Selmi GA, da Cruz AG, Rodrigues CEC, Favaro-Trindade CS (2014) Coencapsulation of xylitol and menthol by double emulsion followed by complex coacervation and microcapsule application in chewing gum. Food Res Int 66:454–462. https://doi.org/10.1016/j.foodres.2014.10.010
7. Holkem AT, Favaro-Trindade CS (2020) Potential of solid lipid microparticles covered by the protein-polysaccharide complex for protection of probiotics and proanthocyanidin-rich cinnamon extract. Food Res Int 136:109520. https://doi.org/10.1016/j.foodres.2020.109520
8. Nori MP, Favaro-Trindade CS, Alencar SM, Thomazini M, Balieiro JCC, Castillo CJC (2011) Microencapsulation of propolis extract by complex coacervation. LWT Food Sci Technol 44:429–435. https://doi.org/10.1016/j.lwt.2010.09.010
9. Souza VB, Thomazini M, Chaves IE, Ferro-Furtado R, Favaro-Trindade CS (2020) Microencapsulation by complex coacervation as a tool to protect bioactive compounds and to reduce astringency and strong flavor of vegetable extracts. Food Hydrocoll 98:105244. https://doi.org/10.1016/j.foodhyd.2019.105244
10. Silva MP, Tulini FL, Matos-Jr FE, Oliveira MG, Thomazini M, Fávaro-Trindade CS (2018) Application of spray chilling and electrostatic interaction to produce lipid microparticles loaded with probiotics as an alternative to improve resistance under stress conditions. Food Hydrocoll 83:109–117
11. Gomez-Estaca J, Comunian TA, Montero P, Furtado RF, Favaro-Trindade CS (2016) Encapsulation of an astaxanthin-containing lipid extract from shrimp waste by complex coacervation using a novel gelatin-cashew gum complex. Food Hydrocoll 61:155–162. https://doi.org/10.1016/j.foodhyd.2016.05.005
12. Comunian TA, Chaves IE, Thomazini M, Moraes ICF, Ferro-Furtado R, Castro IA, Favaro-Trindade CS (2017) Development of functional yogurt containing free and encapsulated echium oil, phytosterol and sinapic acid. Food Chem 237:948–956
13. Xuan J, Xia Q, Tu Y, Luo T, Mao Q, Han Z, Barrow CJ, Liu S, Wang B (2023) Effect of enzymatically produced tuna oil acylglycerol on the characteristics of gelatin O/W emulsion during microencapsulation using complex coacervation. LWT 190:115580
14. Yin M, Chen M, Yuan Y, Liu F, Zhong F (2024) Encapsulation of lactobacillus rhamnosus GG in whey protein isolate-shortening oil and gum Arabic by complex coacervation: enhanced the viability of probiotics during spray drying and storage. Food Hydrocoll 146:109252
15. Comunian TA, Nogueira M, Scolaro B, Thomazini M, Ferro-Furtado R, de Castro IA, Favaro-Trindade CS (2018) Enhancing stability of echium seed oil and beta-sitosterol by their coencapsulation by complex coacervation using different combinations of wall materials and crosslinkers. Food Chem 252:277–284
16. Silva MP, Martelli-Tosi M, Massarioli AP, Melo PS, Alencar SM, Favaro-Trindade CS (2022) Co-encapsulation of guaraná extracts and probiotics increases probiotic survivability and simultaneously delivers bioactive compounds in simulated gastrointestinal fluids. LWT 161:113351
17. Comunian TA, Archut A, Gomez-Mascaraque LG, Brodkorb A, Drusch S (2022) The type of

gum arabic affects interactions with soluble pea protein in complex coacervation. Carbohydr Polym 295:119851

18. Djihad N, Naima FO, Petronilho S, Hamid S, Bedjou FNE, Coimbra MA (2024) Microencapsulation of Citrus limon essential oil by complex coacervation and release behavior of terpenic and derived volatile compounds. Food Hydrocoll 152:109830
19. Soares BS, Constantino ABT, Garcias-Rojas EE (2024) Microencapsulation of curcumin by complex coacervation of lactoferrin and carboxylmethyl tara gum for incorporation into edible films. Food Hydrocoll Health 5:100178
20. Rodriguez MDA, Vidotto DC, Xavier AAO, Mantovani RA, Tavares GM (2023) Does the protein structure of beta-lactoglobulin impact its complex coacervation with type A gelatin and the ability of the complexes to entrap lutein? Food Hydrocoll 140:108651
21. Mu J, Hu R, Tang Y, Dong W, Zhang Z (2024) Microencapsulation of green coffee oil by complex coacervation of soy protein isolate, sodium casinate and polysaccharides: physicochemical properties, structural characterisation, and oxidation stability. Int J Biol Macromol 256: 128064
22. Glomm WR, Molesworth PP, Yesiltas B, Jacobsen C, Johnsen H (2023) Encapsulation of salmon oil using complex coacervation: probing the effect of gum acacia on interfacial tension, coacervation and oxidative stability. Food Hydrocoll 140:108598
23. Liu Y, Jiang J (2023) Preparation of beta-ionone microcapsules by gelatin/pectin complex coacervation. Carbohydr Polym 312: 120839
24. Oliveira AC, Moretti T, Boschini C, Balieiro JCC, Freitas O, Favaro-Trindade CS (2007) Stability of microencapsulated B. lactis (BI 01) and L. acidophilus (LAC 4) by complex coacervation followed by spray drying. J Microencapsul 24:685–693. https://doi.org/10.1080/02652040701532908
25. Archut A, Klost M, Drusch S, Kastner H (2023) Complex coacervation of pea protein and pectin: contribution of different protein fractions to turbidity. Food Hydrocoll 134: 108032
26. Archut A, Drusch S, Kastner H (2022) Complex coacervation of pea protein and pectin: effect of degree and pattern of free carboxyl groups on biopolymer interaction. Food Hydrocoll 133:107884
27. Zou W, Mourad FK, Zhang X, Ahn DU, Cai Z, Jin Y (2020) Phase separation behavior and characterization of ovalbumin and propylene glycol alginate complex coacervates. Food Hydrocoll 108:105978
28. Rabelo RS, Tavares GM, Prata AS, Hubinger MD (2019) Complexation of chitosan with gum arabic, sodium alginate and k-carrageenan: effect of pH, polymer ratio and salt concentration. Carbohydr Polym 223:115120
29. Stounbjerg L, Vestergaard C, Andreasen B, Ipsen R (2019) Associative phase separation of potato protein and anionic polysaccharides. Colloids Surf A 566:104–112
30. Behrouzain F, Razavi SMA, Joyner H (2020) Mechanisms of whey protein isolate interaction with basil seed gum: influence of pH and protein-polysaccharide ratio. Carbohydr Polym 232:115775
31. Vuillemin ME, Michaux F, Muniglia L, Linder M, Jasniewski J (2019) Gum arabic and chitosan self-assembly: thermodynamic and mechanism aspects. Food Hydrocoll 96: 463–474
32. Chen G, Dong S, Chen Y, Gao Y, Zhang Z, Li S, Chen Y (2020) Complex coacervation of zein-chitosan via atmospheric cold plasma treatment: improvement of encapsulation efficiency and dispersion stability. Food Hydrocoll 107:105943
33. Zhang Q, Dong H, Gao J, Chen L, Vasanthan T (2020) Field pea protein/chitosan complex coacervates: formation and characterization. Carbohydr Polym 250:116925
34. Xiong W, Ren C, Tian M, Yang X, Li J, Li B (2017) Complex coacervation of ovalbumin-carboxymethylcellulose assessed by isothermal titration calorimeter and rheology: effect of ionic strength and charge density of polysaccharide. Food Hydrocoll 73:41–50
35. Vijayan UK, Shah NN, Muley AB, Singhal RS (2021) Complexation of curcurmin using proteins to enhance aqueous solubility and bioaccessibility: pea protein vis-à-vis whey protein. J Food Eng 292:110258
36. Perez AA, Carrara CR, Sánchez CC, Rodríguez Patino JM, Santiago LG (2009) Interactions between milk whey protein and polysaccharide in solution. Food Chem 116(1):104–113
37. Ying M, Huang F, Ye H, Xu H, Shen L, Huan T, Huang S, Xie J, Tian S, Hu Z, He Z, Lu J, Zhou K (2015) Study on interaction between curcumin and pepsin by spectroscopic and docking methods. Int J Biol Macromol 79: 201–208
38. Timilsena YP, Wang B, Adhikari R, Adhikari B (2017) Advances in microencapsulation of polyunsaturated fatty acids (PUFAs)-rich

plant oils using complex coacervation: A review. Food Hydrocoll 69:369–381
39. Yan M (2015) Chapter 12: Complex coacervation. In: Mishra M (ed) Handbook of encapsulation and controlled release. Imprint CRC Press, Boca Raton. https://doi.org/10.1201/b19038
40. Santos MG, Bozza FT, Thomazini M, Favaro-Trindade CS (2015) Microencapsulation of xylitol by double emulsion followed by complex coacervation. Food Chem 171:32–39
41. Souza CFJ, Comunian TA, Kasemodel MGC, Favaro-Trindade CS (2019) Microencapsulation of lactase by W/O/W emulsion followed by complex coacervation: effects of enzyme source, addition of potassium and core to shell ratio on encapsulation efficiency, stability and kinetics of release. Food Res Int 121:754–764. https://doi.org/10.1016/j.foodres.2018.12.053
42. Souza VB, Thomazini M, Echalar-Barrientos MA, Nalin CM, Furtado RF, Genovese MI, Favaro-Trindade CS (2018) Functional properties and encapsulation of a proanthocyanidin-rich cinnamon extract (Cinnamomum zeylanicum) by complex coacervation using gelatin and different polysaccharides. Food Hydrocoll 77:297–306. https://doi.org/10.1016/j.foodhyd.2017.09.040
43. Gheonea I, Aprodu I, Cîrciumaru A, Râpeanu G, Bahrim GE, Stănciuc N (2021) Microencapsulation of lycopene from tomatoes peels by complex coacervation and freeze-drying: evidences on phytochemical profile, stability and food applications. J Food Eng 288:110166
44. Vergara LP, dos Santos Hackbart HC, Alves CJ, Reissig GN, Wachholz BS, Borges CD et al (2023) Encapsulation of phenolic compounds through the complex coacervation technique for the enrichment of diet chewable candies. Food Biosci 51:102256
45. Amador C, de Juan LM (2016) Strategies for structured particulate systems design. In: Computer aided chemical engineering, vol 39. Elsevier, pp 509–579
46. You G, Liu XL, Zhao MM (2018) Preparation and characterization of hsian-tsao gum and chitosan complex coacervates. Food Hydrocoll 74:255–266

Chapter 2

Heteroprotein Complex Coacervation: Mechanisms and Encapsulation of Bioactive Compounds

Álvaro Gustavo Ferreira da Silva, Martin Daniel Aya Rodriguez, and Guilherme Miranda Tavares

Abstract

Heteroprotein complex coacervation (HPCC) is a phenomenon of liquid–liquid phase separation in which oppositely charged proteins interact to form a coacervate phase (rich in proteins) that will be in equilibrium with a diluted phase, known as equilibrium solution. Among the different applications of the HPCC, the encapsulation of bioactive compounds has gained special attention due to the potential of these coacervates to protect and improve the bioaccessibility of the target compounds. Nonetheless, several parameters, such as pH, ionic strength, and concentration of the involved molecules, play an important role in the mechanism of complex coacervation. In this context, this chapter will discuss the main methods applied to study the potential of heteroprotein complex coacervates as a vehicle for encapsulating bioactive compounds. The different methods were divided into three segments: (i) characterization of the molecular interaction between the target molecules, (ii) micro and mesoscopic aspects of the formation of complex coacervates, and (iii) coacervation yield and encapsulation efficiency. The principle of the different methods, the experimental protocol, and some insights for data processing will be detailed.

Key words Complex coacervation, Protein complexation, Heteroprotein, Molecular interactions

1 Introduction

Coacervation is an equilibrium phenomenon of a colloidal system characterized by the formation of two distinct liquid phases: one with higher colloid concentration, known as the coacervate phase, and another containing lower colloid concentration, known as equilibrium solution [1]. Coacervates may be formed by a single colloid; and in this case the coacervation is often triggered by water deficit due to the addition of salts or organic solvents [2]. In contrast, when the phase separation is induced by the interaction between colloids displaying opposite charges, the phenomenon is called complex coacervation [1].

Samantha Cristina de Pinho and Carmen S.F. Trindade (eds.), *Bioactives Encapsulation: Food Applications*,
Methods and Protocols in Food Science, https://doi.org/10.1007/978-1-0716-4538-3_2,

Complex coacervation between proteins and polysaccharides has been extensively studied [3–5]; however, complex coacervation between different proteins, known as heteroprotein complex coacervation (HPCC), is considerably less described in the literature [6]. Several intrinsic and extrinsic factors affect the HPCC. The electrostatic-driven molecular interactions are the main triggers of complex coacervation; however, van der Waals forces and hydrophobic interactions are also frequently implicated in the formation of these supramolecular structures. The nature, size, and conformation of the proteins involved, as well as the total protein concentration, ionic strength, and pH are critical parameters modulating the HPCC. Due to the diversity and complexity of the factors affecting the HPCC, characterizing the involved mechanisms in this process is a challenging task.

Similar to other complex coacervation systems, the beginning of HPCC is also governed by electrostatic interactions, and therefore it is highly affected by increasing ionic strength [7]. Typically, HPCC between globular proteins occurs at weak ionic strength and in a narrow pH range, between the pI of the proteins involved [6]. The mixing ratio between proteins is also a critical parameter, in particular because the formation of the elementary supramolecular units for HPCC between globular proteins depends on charge compensation, and also size compensation between the involved proteins [8].

HPCC has attracted attention mainly due to the potential applications of the obtained coacervates, such as for the encapsulation and protection of bioactive compounds [9]. Despite that, only few studies have in depth described the mechanisms involved in the use of complex coacervates formed by different proteins as encapsulation agents [10–15].

The molecular interaction between proteins and ligands often triggers changes in the protein structure, thus the presence of the compound to be encapsulated will impact on the HPCC mechanisms. In this context, the methods applied to characterize the potential of the HPCC for encapsulating bioactive compounds may be grouped into three segments: (i) characterization of the molecular interaction, (ii) micro and mesoscopic aspects of the formation of complex coacervates, and (iii) coacervation yield and encapsulation efficiency. Figure 1 summarizes relevant aspects of the different analytical methods applied for characterizing the encapsulation of target compounds by HPCC.

Regarding the interaction at the molecular level, most of the studies focus on elucidating the mechanisms of interaction between the proteins involved in the HPCC, but they also often address the characterization of the molecular interaction between one of the proteins and the compound that is intended to encapsulate. For these purposes, the Isothermal Titration Calorimetry is usually applied. Relevant information about the elementary bricks that form the complex coacervates can be obtained from the

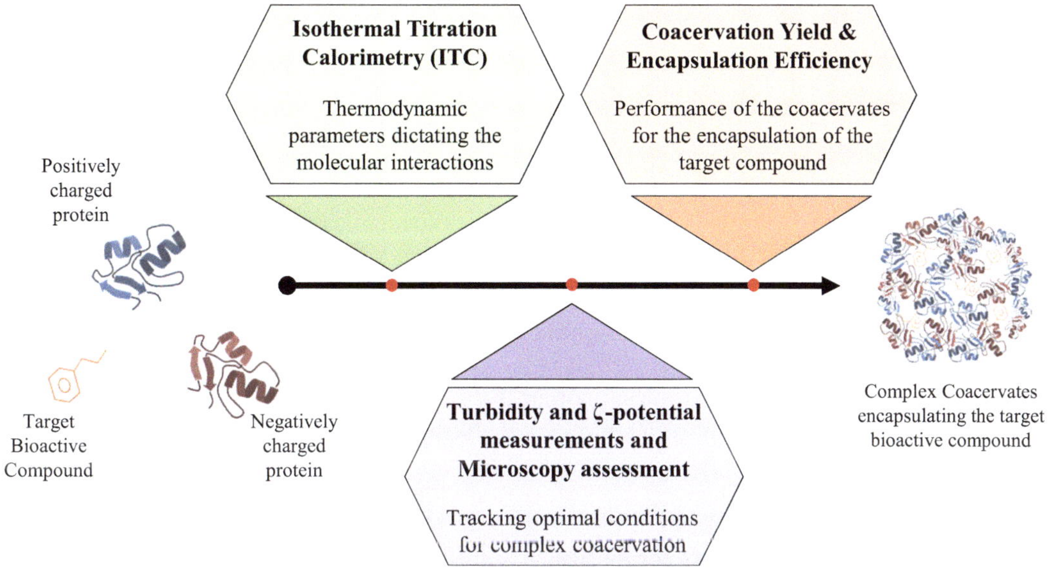

Fig. 1 Relevant aspects of the different analytical methods applied for characterizing the encapsulation of bioactive compounds by heteroprotein complex coacervates

thermodynamic parameters that govern the interaction between the different proteins. Additionally, understanding the features of the molecular interaction between the proteins and the compound to be encapsulated also contributes to better design of the encapsulation process.

A critical issue related to the application of HPCC for the encapsulation of bioactive compounds is to identify the reaction parameters where the coacervation regime is achieved in the (usually) ternary system, formed by the two proteins and the molecule to be encapsulated. In addition to the pH and ionic strength, changes in the concentration and proportion of the involved molecules may induce transition of the complex coacervation regime (liquid–liquid equilibrium) toward the formation of amorphous aggregates, characteristic of liquid–solid equilibrium [11]. To identify the reaction parameters allowing the complex coacervation, turbidity measurements, ζ-potential evaluation, and microscopy assessments are often applied. Once the reaction parameters allowing the complex coacervation are defined, to assess the performance of the encapsulation process the measurements of coacervation yield and encapsulation efficiency are usually monitored. The coacervation yield represents the proportion of proteins into the coacervate phase compared to those remaining in the equilibrium solution. Moreover, the encapsulation efficiency monitors the entrapment of the target compound by the coacervate phase.

This chapter will discuss the principle, experiment protocol and some insights regarding the data processing of these different methods when applied to characterize the HPCC for encapsulating bioactive compounds.

2 Materials

2.1 Characterization of the Molecular Interaction by Isothermal Titration Calorimetry

(a) Samples: protein dispersions and/or bioactive compound solution (*see* **Note 1**).

(b) Dilution buffer solution (*see* **Note 2**).

(c) Hamilton syringe.

(d) Micro or Nanocalorimeter (VP-ITC, Malvern Panalytical, for instance).

(e) Vacuum sample degasser.

2.2 Micro and Mesoscopic Aspects of the Complex Coacervation by Turbidity Measurements

(a) Samples: protein dispersions (*see* **Note 3**) and bioactive compound solution (*see* **Note 4**).

(b) Cuvettes or microplates.

(c) Spectrophotometer or microplate reader.

(d) Micropipettes.

2.3 Micro and Mesoscopic Aspects of the Complex Coacervation by ζ-Potential Measurements

(a) Equipment for ζ-potential measurements (Zetasizer, Malvern Panalytical, for instance).

(b) Samples: protein dispersions and/or bioactive compound solution.

(c) Electrophoretic mobility cell.

(d) Syringe filters and/or centrifuge.

2.4 Micro and Mesoscopic Aspects of the Complex Coacervation by Microscopy Assessment

(a) Optical microscope.

(b) Sample: protein dispersions and/or bioactive compound solution.

(c) Slides and coverslips.

2.5 Encapsulation Performance by Coacervation Yield and Encapsulation Efficiency Measurements

(a) Samples: protein dispersions and bioactive compound solution.

(b) Microtubes.

(c) Centrifuge.

(d) Micropipettes.

(e) Required materials for the quantification of the involved proteins and target bioactive compound.

3 Methods

3.1 Characterization of the Molecular Interaction by Isothermal Titration Calorimetry (ITC)

Isothermal titration calorimetry (ITC) allows studying the thermodynamic parameters that dictate the molecular interaction between two different proteins or between a protein and a ligand. In the context of encapsulation of bioactive compounds by HPCC, the ITC can be applied to (i) study the formation of the elementary bricks of complex coacervation between the two involved proteins and/or (ii) understand the affinity between the target bioactive compound and each of the proteins involved. Collecting these data is useful for designing the experiments to track the optimum conditions for coacervation and encapsulation.

The ITC equipment has an automated syringe that performs successive injections of the solution/dispersion of one of the molecules into an adiabatic cell containing the solution/dispersion of the other molecule under study (Fig. 2a). The experiment is carried out at a constant temperature; and after each injection, the heat released or consumed during the molecular interaction is monitored (Fig. 2b). Therefore, this technique allows the study of molecular interactions displaying positive or negative change in the enthalpy. Since the consumption or release of heat is proportional to the intensity of formation of the molecular complexes, the isotherms obtained at the end of the titration by integrating the signal acquired from each injection allow finding the association constant (Ka), the stoichiometry (n), and the change in enthalpy (ΔH) related to the interaction (Fig. 2c). These results are generated by fitting the experimental data to mathematical models. Moreover, from these parameters the change in Gibbs free energy (ΔG) and the change in the entropy (ΔS) may be deducted by applying Eqs. (1) and (2).

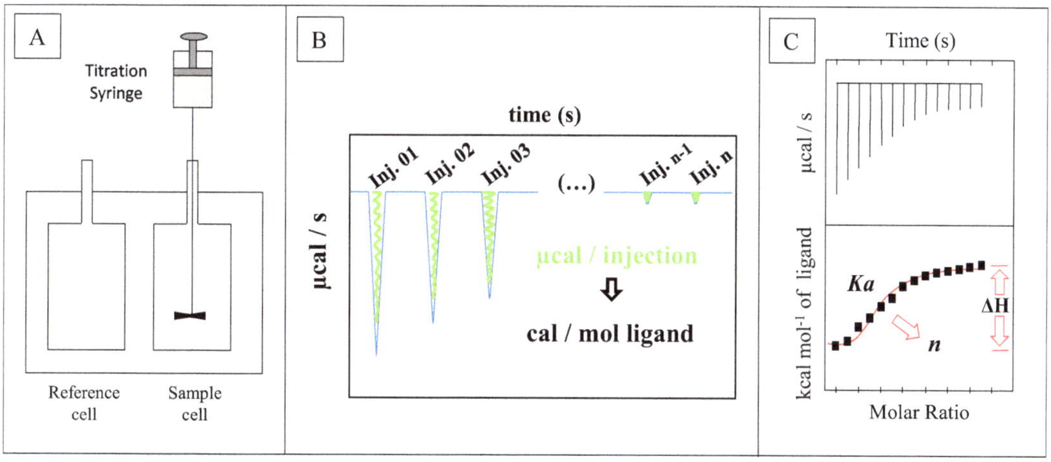

Fig. 2 Schematic representation of the isothermal titration calorimetry (ITC) device (**a**), the titration progress (**b**), and the fit of the experimental data to obtain the thermodynamic parameters of the interaction (**c**)

$$\Delta G = -R \times T \times \ln(\mathrm{Ka}) \tag{1}$$

$$\Delta G = \Delta H - \mathrm{T} \times \Delta S \tag{2}$$

where R is the constant of ideal gases (J.mol^{-1}.K^{-1} or kcal.mol^{-1}.K^{-1}); T is the temperature (K); Ka is the association constant (M^{-1}); ΔG is the change in the Gibbs free energy (J.mol^{-1} or kcal.mol^{-1}); ΔH is the change in the enthalpy (J.mol^{-1} or kcal.mol^{-1}); and ΔS is the change in the entropy (J.mol^{-1}.K^{-1} or kcal.mol^{-1}.K^{-1}).

The method for ITC measurements can be performed as follows:

(a) The setup of several parameters before analysis is critical. Among these critical parameters, the following deserve special attention:

(a.1) Temperature: the analysis is conducted at a fixed temperature. Monitoring thermodynamic parameters of the interaction at different temperatures allows obtaining additional information on the interaction such as heat capacity change (ΔCp).

(a.2) Initial delay: corresponds to the time between the beginning of the analysis and the first injection. Usually, it is not lower than 60 s.

(a.3) Stirring speed: corresponds to the stirring speed in the sample cell. It is crucial to ensure homogeneity of the sample cell content after each injection. It is often set at 300 rpm.

(a.4) Volume of each injection and total number of injections: these parameters are at the heart of the experimental design. Together with the concentration of the solution/dispersion in the titration syringe and the one into the sample cell (*see* **Note 1**), the volume of each injection and the total number of injections will determine the molar ratio between the studied molecules at the end of the titration. At the end of the titration, it is expected that the molar ratio between the molecules will allow the interaction to be close to saturation (*see* **Note 5**).

(a.5) Duration of each injection: the set time depends on the characteristics of the dispersion/solution in the syringe (such as its viscosity) and the volume of each injection. Generally, it is set between 10 and 20 s.

(a.6) Time between successive injections: the time between two successive injections needs to allow a reestablishment of the baseline after the signal produced by each injection. It is generally established between 200 and 300 s.

(b) Place into the appropriate vessels the solution/dispersion that will go into the syringe, the solution/dispersion that will be allocated into the sample cell, and the dilution buffer that will go into the reference cell. Using the vacuum sample degasser, equilibrate the solutions/dispersions at the working temperature and degas them (*see* **Note 6**).

(c) Using the Hamilton syringe, fill the reference cell and the sample cell with the respective solutions/dispersions. Ensure that the cells are perfectly filled by removing the excess of the solutions/dispersions.

(d) Fill the titration syringe with the respective solution/dispersion.

(e) Carefully place the titration syringe into the sample cell.

(f) Wait for the system to reach working temperature and start the analysis (*see* **Note 7**).

(g) Repeat the procedure to perform the control titrations: (i) Titrate the syringe solution/dispersions into the sample cell containing only the dilution buffer and (ii) fill up the titration syringe with the dilution buffer and titrate it into the sample cell containing the sample solution/dispersion.

3.1.1 Insights for Data Treatment

Figure 3a displays a classic representation of an isotherm obtained for a molecular interaction featuring a single class of binding site. By fitting the experimental data with this binding model, the thermodynamic parameters of the interaction can be easily

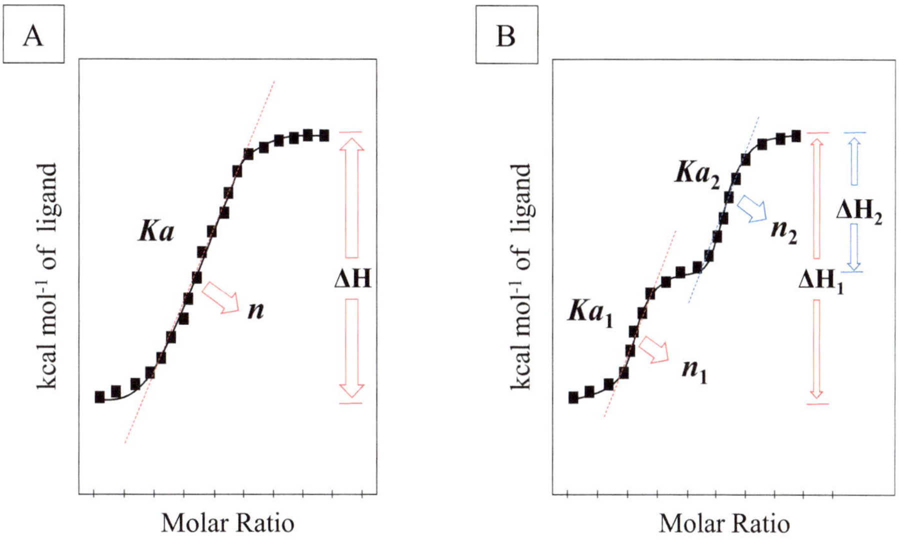

Fig. 3 Representation of an isotherm obtained for a molecular interaction featuring a single class of binding site (**a**) and representation of an isotherm obtained for a molecular interaction featuring two classes of binding sites (**b**)

obtained. This type of isotherm is frequently described either for interactions between oppositely charged proteins and between proteins and bioactive compounds [15]. In turn, Fig. 3b brings a generic representation of an isotherm obtained for a molecular interaction having two different classes of binding sites. Likewise, through the adjustment of pertinent mathematical models, the thermodynamic parameters associated to each class of binding sites may be obtained. For instance, isotherms displaying two classes of binding sites were described for the molecular interaction between lactoferrin and β-lactoglobulin [16]. More complex isotherm profiles have also been reported in the literature and they are often associated with the overlap of different events such as molecular interaction followed by protein aggregation [17].

Additionally, the isotherms obtained from control titrations need to be carefully evaluated during data processing. When only a weak signal associated with the dilution of the solutions/dispersions is obtained, it can be easily deduced from the one obtained during the molecular interaction before data processing. However, isotherms with specific characteristics may be obtained in control titrations depending on the involved molecules. Dissociation of dimeric proteins or release of metals from metal-binding proteins can produce isotherms with specific profiles [16, 18]. In these cases, it is important to evaluate whether or not it makes sense to deduce the signal linked to these events from the one obtained for the molecular interaction.

3.2 Micro and Mesoscopic Aspects of the Complex Coacervation by Turbidity Measurements

Turbidimetry is a technique based on monitoring the reduction of the transmittance of the light that is incident on a sample. The measurement is made spectrophotometrically in the visible region of the electromagnetic spectrum, usually at 600 nm. This technique is widely applied to study the potential of HPCC to encapsulate bioactive compounds, as it allows monitoring the formation of supramolecular structures resulting from the interaction between the proteins, in the presence or absence of the bioactive compound. By monitoring the increase in turbidity, it is possible to identify the pH range and mixing proportions allowing the formation of molecular complexes, as well as the kinetics of formation and growth of such supramolecular structures [19, 20].

(a) Mix the right amount of protein dispersions or protein dispersions and bioactive compound solution (*see* **Note 8**) in the cuvette or microplate well (*see* **Note 9**).

(b) Read the absorbance of the resulting mixture at 600 nm.

(c) Turbidity is calculated by $\tau = (2.303 \times A_{600nm})/l$, where τ is the turbidity (cm^{-1}), A_{600nm} is the absorbance at 600 nm, and l is the optical path (*see* **Note 10**).

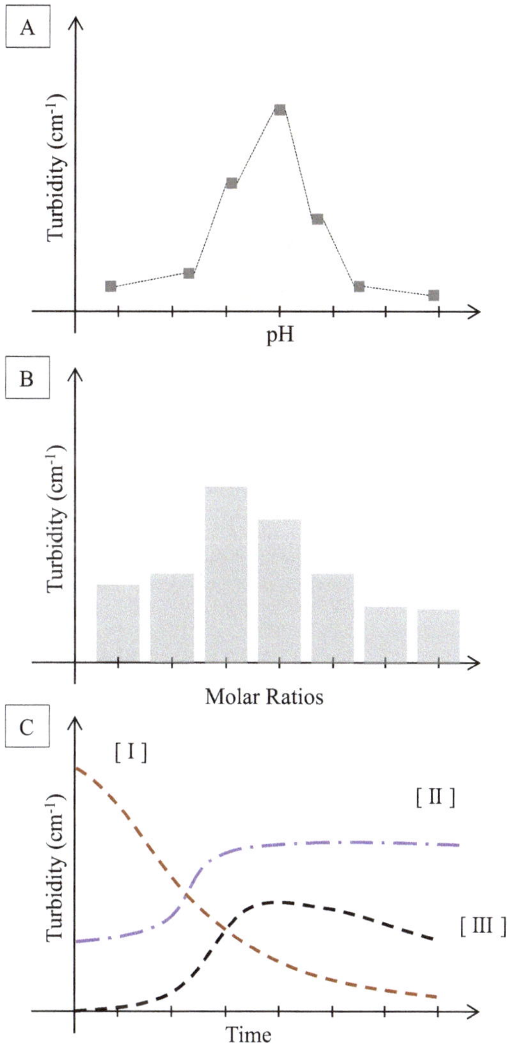

Fig. 4 Generic representation of the turbidity measurements of heteroprotein complex coacervation (HPCC) systems as a function of pH (**a**). Molar ratio between the involved molecules (**b**) and time (**c**)

3.2.1 Insights for Data Treatment

Figure 4 displays generic representations of turbidity measurements applied to track reactional parameters allowing the encapsulation of bioactive compounds by HPCC. Figure 4a exemplifies the application of turbidity measurements to identify pH ranges where there is intense formation of supramolecular structures in binary (containing the oppositely charged proteins) or ternary systems (containing the bioactive compounds in addition to the oppositely charged proteins) [10, 16]. In this representation, the highlighted zone, where higher turbidity is observed, represents the pH range that favors the interaction between the involved molecules (Fig. 4a). Similarly, Fig. 4b represents the application of turbidity measurements to monitor at which mixing molar ratios there is a prevalence of supramolecular structures formation. Once again, higher turbidities are consistent with greater formation of supramolecular structures. In this specific case, Fig. 4b only brings information for a fixed pH value. To cover a broader pH range, the different molar ratios should be evaluated at different pH values (Fig. 4a, b).

Taking into account the turbidity measures alone, it is difficult to guarantee that complex coacervates have been formed (liquid–liquid equilibrium), since changes in turbidity are also observed when amorphous aggregates are formed as the result of the molecular interaction. Usually, microscopy assessments are applied to confirm that complex coacervates have been formed (*see* Subheading 3.4).

In complement, additional information regarding the formed supramolecular structures may be obtained by monitoring the turbidity kinetics (Fig. 4c). Great variability in the resulting profile may be observed as a function of the nature of the involved molecules. The curve [I] in Fig. 4c generically represents the turbidity kinetics observed during the encapsulation of different molecules by lactoferrin and β-lactoglobulin complex coacervates [10, 11]. In this curve, there is quick formation of complex coacervates immediately after mixing the different molecules and then there is a reduction in turbidity as a result of the coalescence of the complex coacervates. Additionally, curve [II] in Fig. 4c illustrates the turbidity kinetics where amorphous aggregates were observed, for this same system. Under these conditions, there was an initial formation of small aggregates, which grew over time [10, 11]. Moreover, curve [III] in Fig. 4c displays characteristic turbidity kinetics where complex coacervates were observed; nevertheless, under the tested conditions, no turbidity changes were noticed after mixing the different molecules; however, over time there was the formation of complex coacervates, which eventually coalesced [15]. Nevertheless, it is important to state that no universal behaviors have been described, since they are quite dependent on the target molecules and reaction parameters.

3.3 Micro and Mesoscopic Aspects of Complex Coacervation by ζ-Potential Measurements

The ζ-potential measurements are widely applied to obtain information about the magnitude of the electrostatic repulsion or attraction between macromolecules. In the context of HPCC, since electrostatic interaction is the main driving force behind the formation of the complex coacervates building blocks, ζ-potential measurements are almost mandatory to easily track the optimal conditions allowing coacervation to occur. Charged particles suspended in a liquid medium are surrounded by a double layer: (i) a liquid layer closer and strongly linked to the particle, known as Stern layer and (ii) another continuous layer, weakly interacting

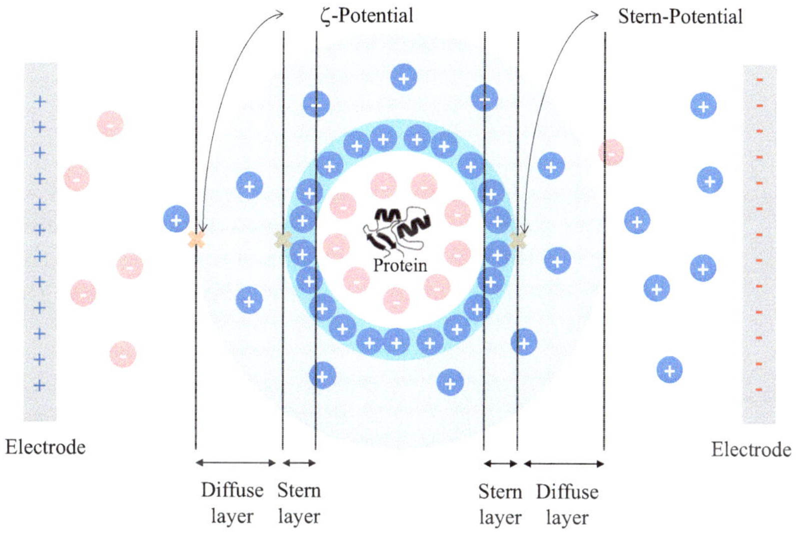

Fig. 5 Representation of the Stern and diffuse layers surrounding a protein suspended in a liquid medium

with the particle, known as diffuse layer. Stern potential is the electric potential at the outer limit of the Stern layer, where there is greater charge density. On the other hand, ζ-potential is the electric potential measured at the outer limit of the diffuse layer, where there is a lower charge density. Charged particles can be studied by applying an electric field to the liquid and monitoring their mobility toward the electrodes according to their affinity (Fig. 5) [21, 22].

The HPCC occurs at pH between the isoelectric points of the involved proteins. When applied to encapsulate bioactive compounds, the molecular interaction between the proteins and the target compound may influence the measured ζ-potential. Thus, ζ-potential measurements are useful not only to track the pH of complex coacervation occurrence but also to investigate mixing molar ratios allowing it. Unlike the complex coacervation between other macromolecules, the HPCC between globular proteins has been described as affected not only by charge compensation, but also by size compensation [8]. Laser Doppler electrophoresis is the most used method to verify the ζ-potential and consists of measuring the speed of movement of particles in a liquid exposed to an electric field [19].

(a) Allocate the individual dispersions or their mixtures into the electrophoretic mobility cells (*see* **Notes 11** and **12**).

(b) Equilibrate the samples in the measurement temperature.

(c) Apply an electric potential of 150 V, for instance, to induce electrophoretic mobility.

(d) The ζ-potential may be calculated from the measured electrophoretic mobility by applying the Henry's equation:

$$\frac{U}{E} = \frac{2\varepsilon\,\zeta\,F(\mathrm{Ka})}{3\eta}$$

where U/E is the electrophoretic mobility ($\mathrm{m}^2\mathrm{V}^{-1}\mathrm{s}^{-1} \times 10^{-8}$); ε is the dielectric constant of the continuous medium (dimensionless); ζ is the ζ-potential (mV); η is the viscosity of the continuous medium (*cP*); and *F*(Ka) is the Henry function (*see* **Notes 13** and **14**).

3.3.1 Insights for Data Treatment

ζ-potential data is usually plotted against medium pH or mixing ratio when applied to study the potential of HPCC for encapsulating target compounds, as generically represented in Fig. 6. Usually, for binary or ternary systems ζ-potential values close to zero indicate greater complexation between the involved molecules. Thus, these data complement information extracted from turbidity measurements.

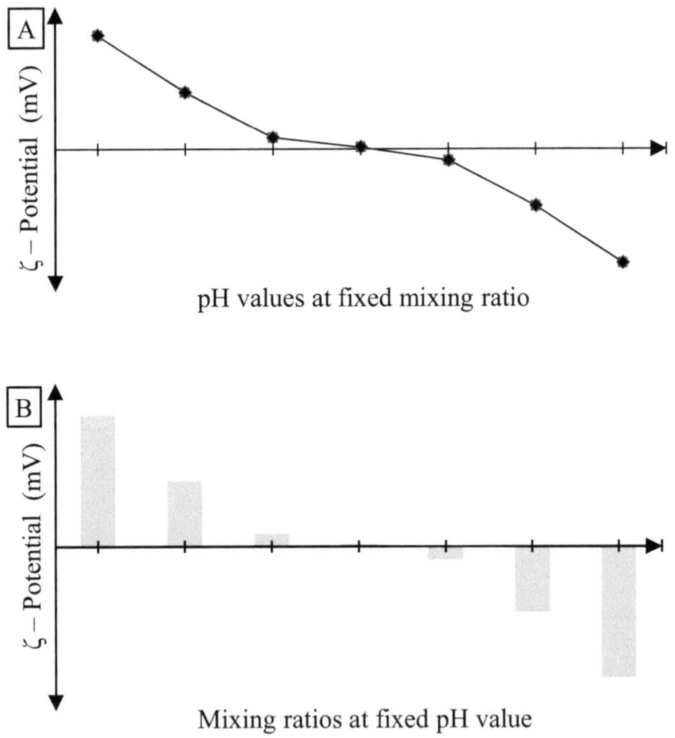

Fig. 6 Generic representation of ζ-potential measurements as a function of pH values (**a**) or mixing ratios (**b**) for tracking reaction conditions allowing the HPCC

3.4 Micro and Mesoscopic Aspects of the Complex Coacervation by Microscopy Assessment

Optical microscopy is a very simple and useful tool for identifying reaction conditions allowing the HPCC. In this context, the bright field microscopy is mostly used to distinguish coacervates from amorphous aggregates. Supramolecular structures whose shape remains unidentifiable by bright field microscopy even after adjusting the illumination can be subjected to complementary optical microscopy techniques, such as dark-field microscopy and confocal laser microscopy [19, 23].

(a) Mix the right amount of protein dispersions and/or bioactive compound solution in the microtubes.

(b) Deposit an aliquot of the sample on the slide and place the coverslip (*see* **Note 15**).

(c) Place the slide containing the sample on the microscope.

(d) Select the ideal objective and adjust the focus (*see* **Note 16**).

(e) Readjust lighting, if necessary for data acquisition (*see* **Note 17**).

3.4.1 Insights for Data Treatment

Microscopy assessment is primarily used to distinguish complex coacervates, which generally have a spherical shape (Fig. 7a), from amorphous aggregates (Fig. 7a). This analysis allows characterizing the reaction conditions (pH, protein concentration, and mixing ratio) where complex coacervates are effectively formed [10, 11].

3.5 Encapsulation Performance by Coacervation Yield and Encapsulation Efficiency measurements

Coacervation yield and encapsulation efficiency are also two useful parameters to identify the optimal conditions for encapsulation of bioactive compounds by HPCC. Despite turbidity measurements being applied to track the optimal conditions for complex coacervation (pH, ionic strength, protein concentration, and protein molar ratio), the determination of the coacervation yield is essential

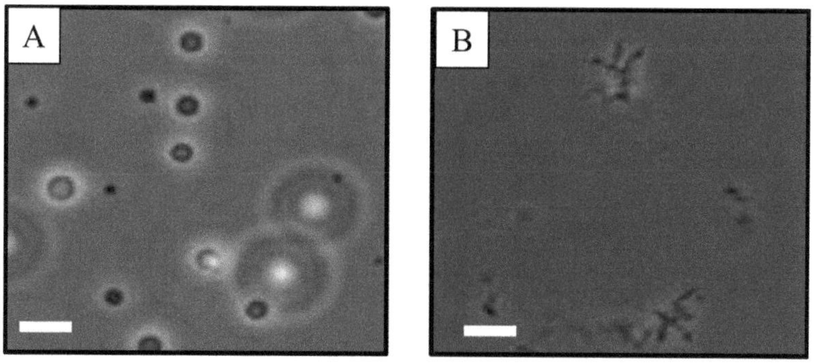

Fig. 7 Optical microscopy of ternary mixtures between lactoferrin, β-lactoglobulin, and folic acid. The transition from complex coacervation (**a**) to amorphous aggregation (**b**) regimes triggered by the increase of folic acid concentration at the same protein concentration [24]. The white bars represent 10 μm

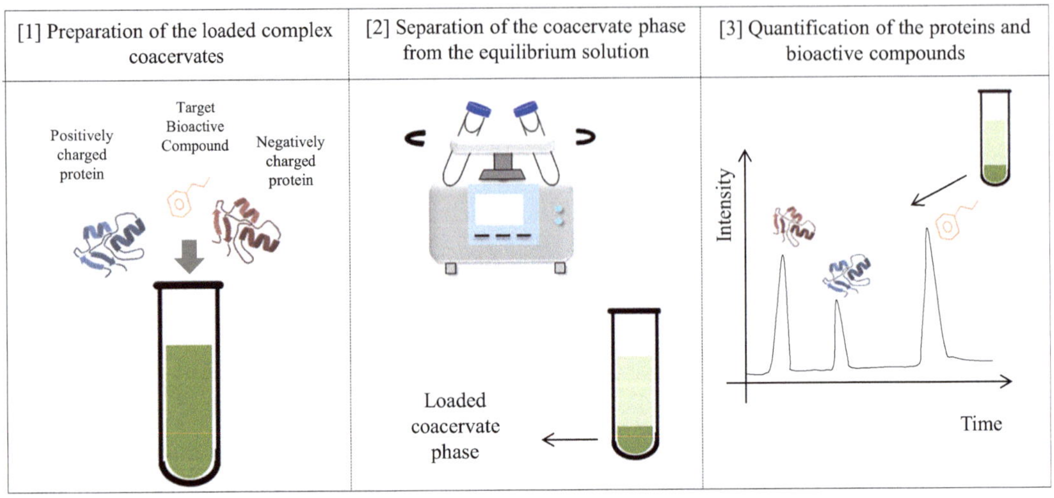

Fig. 8 Steps for determining the coacervation yield and the encapsulation efficiency

for confirming these data. Basically, the coacervation yield corresponds to the proportion of initially mixed proteins that effectively form the coacervate phase. The coacervation yield is unlikely to be 100%, once the mechanism of coacervation involves the formation of an equilibrium solution, displaying low colloid concentration [1]. Likewise, encapsulation efficiency consists of determining the proportion of the initial amount of bioactive compounds mixed with the protein dispersions that is effectively entrapped by the complex coacervates. The determination of both parameters is based on quantifying the proteins and bioactive compounds in the coacervate phase after their separation from the equilibrium solution by centrifugation (Fig. 8).

(a) Mix the right amount of protein dispersions and bioactive compound solution (*see* **Note 18**) in the microtubes.

(b) If necessary, wait the required time for formation of complex coacervates (*see* **Note 19**).

(c) Separate the coacervate phase from the equilibrium solution by centrifugation (at least $15{,}000 \times g$ for 10 min).

(d) Carefully remove the equilibrium solution (lean phase) from the coacervate phase.

(e) Using appropriate methods, quantify the amount of proteins and bioactive compounds into the coacervate phase (*see* **Note 20**). Alternatively, the quantification of proteins and bioactive compounds remaining in the equilibrium solution may be performed, and the amounts composing the coacervate phase can be deducted by difference.

(f) Coacervation yield (CY) can be obtained from the following equation: $\%\ CY = 100 \times (m_{pc}/m_{pi})$, where m_{pc} is the amount of proteins forming the coacervate phase and m_{pi} is the amount of proteins initially mixed. In turn, the encapsulation efficiency (EE) can be calculated by applying the following equation: $\% \ EE = 100 \times (m_{BCc}/m_{BCi})$, where m_{BCc} is the amount of bioactive compounds entrapped by the coacervate phase and m_{BCi} is the amount of bioactive compounds initially mixed with the protein dispersions.

3.5.1 Insights for Data Treatment

In addition to coacervation yield and encapsulation efficiency, the maximum mass ratio between bioactive compounds and proteins in the coacervate phase can be calculated [10]. This parameter is useful for comparing the encapsulation by HPCC with other encapsulation methods applied to the same bioactive compound.

4 Notes

1. The concentration of the protein dispersions and/or bioactive compound solutions will depend on the design of the experiment. Typically, the molar concentration of the solution/dispersion in the syringe is considerably greater than one of the solution/dispersion placed in the sample cell. In this way, during the titration, the concentration of the titrant into the sample cell will increase progressively after each injection, allowing the interaction to approach saturation at the end of the analysis. The extraction of relevant information from the obtained isotherm will depend, among other factors, on the titration to cover molar ratios beyond the interaction stoichiometry and close to its saturation. In the case of the interaction between bioactive compounds and proteins, the protein dispersion usually is allocated in the sample cell, while the bioactive compound solution is placed into the syringe (Conventional Titration). However, depending on the characteristics of the target bioactive compound, it may be challenging to produce stable solutions at the concentrations required for analysis (formation of aggregates, for instance). In these cases, it is recommended to perform the Reverse Titration, with the protein dispersion placed into the syringe and the bioactive compound solution into the sample cell [18, 25]. Essentially, the molecule displaying the lowest solubility must be the one that remains into the sample cell. Regarding the interaction between different proteins, this same rationalization is required. Depending on the nature of the involved proteins, obtaining stable dispersions at the required concentration for the syringe may be critical. Some proteins, such as gelatins, are capable of forming gels at relatively low concentrations; therefore, they must be preferably allocated into the sample cell [15].

2. The chemical compatibility between the involved molecules and the used buffer solution is mandatory. Likewise, the buffer solution used to prepare the solution/dispersion placed into the syringe must match the buffer solution used to prepare the solution/dispersion allocated into the sample cell. The release of counter-ions during titration generates thermal signals; therefore, it is recommended to use buffer solutions that guarantee pH control throughout the titration.
3. The concentration, pH, and ionic strength of protein dispersions are essential aspects of sample preparation. The maximum protein concentration of the dispersion will depend on the solubility of the proteins and/or their gelling properties (as in the case of gelatin, for instance). The presence of protein aggregates may compromise the ability of proteins to interact, affecting the obtained results. Very often, protein dispersions are filtered (0.45 μm pore size filters) after preparation. Dispersions may be prepared in water or buffer solution. In both cases, adjusting the pH is critical. When dispersions are prepared in buffer solution, it is important to ensure compatibility between the buffer and the protein (phosphate buffer should be avoided to disperse casein micelles, for instance). In this circumstance, the concentration of the buffer solution should not be excessive, as the increase in ionic strength will have a negative impact on the interaction between the proteins. However, the use of a buffer solution may ensure pH control even after mixing the protein dispersions. When protein dispersions are prepared in water, the ionic strength is minimal, but the pH can vary after mixing the proteins due to the release of counter-ions during the interaction. Therefore, monitoring the pH will be required.
4. The target bioactive compound solution must be clear and its pH must also be monitored/adjusted (cf. **Note 3**). In case of low solubility of the bioactive compound and the need to use small portions of organic solvent, it is imperative to take into account that the solvent will affect the dynamics of molecular interaction. When this strategy is applied, usually, the concentration of organic solvent does not exceed 5% v/v in the final mixture [15].
5. To understand the thermodynamic parameters that dictate the formation of the elementary bricks of HPCC, it is important that the experimental design focuses on a range of concentrations and molar ratios between the involved molecules that favor the formation of molecular complexes, but restrict their growth (formation of larger supramolecular structures). The occurrence of different events during titration may produce isotherms that are difficult to interpret. At the same time, dispersions/solutions need to be concentrated enough to

produce an exploitable signal during titration. An interesting strategy is to simulate, outside the sample cell, the titration process that will occur during the analysis. This strategy ensures that the working concentrations and molar ratios will not induce gelation of the system (which can happen at relatively low contractions when gelatins are involved in the analysis, for instance).

6. Degassing is a critical step in sample preparation. The formation of air microbubbles during the analysis may compromise the quality of the obtained isotherm due to the excessive background noise generated.
7. Monitor the evolution of the baseline before starting the analysis. To improve the quality of the obtained signal, it is useful to launch the analysis only when a stable baseline is observed.
8. The turbidity measurements may be applied in different ways to evaluate the potential of the HPCC to encapsulate bioactive compounds. This analysis may be used to monitor the formation of supramolecular structures by the proteins both in the absence (binary system) and in the presence (ternary system) of the bioactive compound. To evaluate the pH range where supramolecular structures are formed, generally protein dispersions and/or bioactive compound solution are independently prepared at the same pH values and then mixed. Thus, to cover a broad pH range, different sets of dispersions/solution should be prepared. The increasing or decreasing the pH of the mixture can impact the equilibrium of the interaction between the molecules, and therefore impair the formation of complex coacervates [26]. In HPCC, the formation of coacervates will be observed at a pH value between the pI values of the involved proteins [6]. The effect of the total protein and/or bioactive compound concentration, and the mixing ratio between the involved molecules in the formation of supramolecular structures can also be studied by turbidity measurements. In these circumstances, different volumes of the stock solutions/dispersions should be mixed to achieve the required concentration/proportion. Additionally, to monitor the kinetics of formation of supramolecular structures, the measurement of turbidity over time may also be performed. In the case of ternary systems, the order of mixing often has an impact on the observed results. Therefore, for these systems, usually the bioactive compound is added to one of the proteins and then the other protein is mixed.
9. When preparing the mixtures, it is imperative to avoid the formation of air bubbles, which can affect turbidity measurements.

10. Although dependent on the configurations of the equipment used, the linearity range of absorbance measurements is usually valid for values not exceeding 1. If required, to overcome this issue, cuvettes with a reduced optical path may be used. Regarding the microplates, the optical path often corresponds to the height of the liquid column in the microplate well; therefore, to modulate the optical path, it is possible to modulate the final volume of the mixture.
11. Handle the sample slowly to avoid the formation of bubbles. If bubbles are still formed, degassing the sample may be a useful strategy.
12. Individual dispersions or their mixtures should be sufficiently diluted. To ensure the absence of insoluble suspended material, dispersions/solutions may be centrifuged and/or filtered (e.g., in 0.45 μm pore size syringe filters).
13. The Henry equation is commonly used with the Smoluchowski approximation, where $F(\mathrm{Ka}) = 1.5$.
14. The dielectric constant and the viscosity depend on the continuous phase used for sample preparation. In the case of water, the dielectric constant and viscosity are 78.5 and 0.887 mPa·s, respectively.
15. Samples may be analyzed just after mixing or after a specific incubation time allowing the growth of the supramolecular structures. Additionally, to maximize the number of structures in the analyzed aliquot, samples may be subjected to an additional centrifugation step (e.g., at 15,000×g for 10 min at 25 °C).
16. The total magnification is the result of multiplying the magnification of the objective lens by the magnification of the ocular lens (e.g., using a 10× ocular lens and a 100× objective, there is a total magnification of 1000×).
17. Sample visualization and data acquisition must be performed as quickly as possible, as the incidence of light may heat the sample affecting the observed supramolecular structures.
18. Turbidity measurements may be useful to define the most promising mixing molar ratios, protein and bioactive compound concentrations, mixing order, and pH.
19. Depending on the involved molecules or working conditions, the assembling of the elementary bricks to form the coacervates may take some time, which remains a variable to be optimized case by case [15].
20. Different methods can be used to quantify proteins or bioactive compounds in the coacervate phase. Specific methods involving chromatographic separations have often been applied [10, 12]. In these cases, the development and validation of the method may be required. Spectrophotometric protein

quantification methods can also be applied [15]. In these cases, it is important to take into account that complex coacervates are heterogeneous systems and different proteins may exhibit different responses when subjected to these indirect methods. Concerning the bioactive compounds, their quantification may also involve different extraction steps with organic solvents depending on their chemical nature [13, 15].

References

1. McNaught AD, Wilkinson A (1997) Compendium of chemical terminology, vol 1669. Blackwell Science, Oxford
2. Bachtsi AR, Kiparissides C (1996) Synthesis and release studies of oil-containing poly (vinyl alcohol) microcapsules prepared by coacervation. J Control Release 38(1):49–58
3. Kayitmazer AB et al (2013) Protein–polyelectrolyte interactions. Soft Matter 9(9): 2553–2583
4. Kizilay E, Kayitmazer AB, Dubin PL (2011) Complexation and coacervation of polyelectrolytes with oppositely charged colloids. Adv Colloid Interf Sci 167(1–2):24–37
5. Schmitt C, Turgeon SL (2011) Protein/polysaccharide complexes and coacervates in food systems. Adv Colloid Interf Sci 167(1–2): 63–70
6. Croguennec T, Tavares GM, Bouhallab S (2017) Heteroprotein complex coacervation: a generic process. Adv Colloid Interf Sci 239: 115–126
7. Nigen M, Croguennec T, Bouhallab S (2009) Formation and stability of α-lactalbumin–lysozyme spherical particles: involvement of electrostatic forces. Food Hydrocoll 23(2): 510–518
8. Desfougères Y et al (2010) Charge and size drive spontaneous self-assembly of oppositely charged globular proteins into microspheres. J Phys Chem B 114(12):4138–4144
9. Tavares GM et al (2014) Milk proteins as encapsulation devices and delivery vehicles: applications and trends. Trends Food Sci Technol 37(1):5–20
10. Chapeau A-L et al (2016) Spontaneous co-assembly of lactoferrin and β-lactoglobulin as a promising biocarrier for vitamin B9. Food Hydrocoll 57:280–290
11. Tavares GM et al (2017) How the presence of a small molecule affects the complex coacervation between lactoferrin and β-lactoglobulin. Int J Biol Macromol 102:192–199
12. Diarrassouba F et al (2015) Food protein-based microspheres for increased uptake of vitamin D3. Food Chem 173:1066–1072
13. Zhu J et al (2019) Construction of fucoxanthin vector based on binding of whey protein isolate and its subsequent complex coacervation with lysozyme. J Agric Food Chem 67(10): 2980–2990
14. Zheng J et al (2022) Encapsulation behavior of curcumin in heteroprotein complex coacervates and precipitates fabricated from β-conglycinin and lysozyme. Food Hydrocoll 133:107964
15. Rodriguez MDA et al (2023) Does the protein structure of β-lactoglobulin impact its complex coacervation with type A gelatin and the ability of the complexes to entrap lutein? Food Hydrocoll 140:108651
16. Tavares GM et al (2015) Selective coacervation between lactoferrin and the two isoforms of β-lactoglobulin. Food Hydrocoll 48:238–247
17. Tavares GM et al (2015) Binding of folic acid induces specific self-aggregation of lactoferrin: thermodynamic characterization. Langmuir 31(45):12481–12488
18. Mantovani RA et al (2020) Unraveling the molecular mechanisms underlying interactions between caseins and lutein. Food Res Int 138: 109781
19. Zheng J, Tang C-H, Sun W (2020) Heteroprotein complex coacervation: focus on experimental strategies to investigate structure formation as a function of intrinsic and external physicochemical parameters for food applications. Adv Colloid Interf Sci 284:102268
20. Gu R et al (2024) Effects of catechol grafting on chitosan-based coacervation and adhesion. Int J Biol Macromol 267:131662
21. Cano-Sarmiento C et al (2018) Zeta potential of food matrices. Food Eng Rev 10:113–138
22. Lowry GV et al (2016) Guidance to improve the scientific value of zeta-potential measurements in nanoEHS. Environ Sci Nano 3(5): 953–965

23. Hoffmann K et al (2015) A molecular view of the role of chirality in charge-driven polypeptide complexation. Soft Matter 11(8): 1525–1538
24. Tavares GM (2015) β-lactoglobulin and lactoferrin complex coacervates: Characterization and putative applications as encapsulation device. [Doctoral thesis, Agrocampus Ouest]. Available at: https://hal.science/tel-01686515/
25. Mantovani RA et al (2021) Binding of carotenoids to milk proteins: why and how. Trends Food Sci Technol 110:280–290
26. Yan Y et al (2013) pH-dependent aggregation and disaggregation of native β-lactoglobulin in low salt. Langmuir 29(14):4584–4593

Chapter 3

Ionotropic Gelation

Silvia Cristina Sobottka Rolim de Moura and Izabela Dutra Alvim

Abstract

The ionic gelation technique is widely used in the encapsulation of different types of substances, with applications in different segments, including food. This technique has many advantages such as mild production conditions that help maintain the characteristics of the encapsulated substances. The technique has some limitations such as the retention of hydrophilic substances, which can be overcome by using a double emulsion. This chapter presents the ionic gelation technique for encapsulating substances. The most commonly used wall materials, pectin and alginate, and the main salts to promote polymer cross-linking are described. Practical procedures for obtaining laboratory-scale particles are described. The main analyses for this type of particles are mentioned. Some problems that may occur during the procedure are listed. Various applications of particles obtained in the study are described.

Key words Ionotropic gelation, Polysaccharides, Ingredients encapsulation, Crosslinking

1 Introduction

Ionotropic gelation relies on the ability of various hydrocolloids, especially anionic gums like alginate, pectin, gellan gum, and carrageenan, to form a gel matrix in the presence of divalent cations. The interaction of ions, such as Ca^{+2}, with the carboxyl groups of polysaccharide chains leads to the formation of insoluble gels. When a polysaccharide dispersion containing an active ingredient is dripped into an ion solution, particles are created. After a specific contact time in the ionic bath, these particles reach maximum stability, and their mechanical strength becomes constant. The concentration of calcium also influences the rigidity of the particles [1–3].

For microencapsulation, ionotropic gelation can be achieved through methods like atomization, extrusion and co-extrusion, or electrostatic spraying. This technique has the advantage of using very mild conditions, without requiring high temperatures, vigorous agitation, or organic solvents, which allows for the encapsulation of substances that would degrade under harsher conditions

Samantha Cristina de Pinho and Carmen S.F. Trindade (eds.), *Bioactives Encapsulation: Food Applications*, Methods and Protocols in Food Science, https://doi.org/10.1007/978-1-0716-4538-3_3,

[4]. However, a disadvantage is that encapsulating hydrophilic or low molecular weight materials often results in easy diffusion and rapid release through the ionic gel network, regardless of pH [5]. To effectively retain hydrophilic active compounds, additional strategies like emulsion systems or coating materials are necessary, as ionotropic gelation is primarily suited for hydrophobic or poorly soluble compounds [6–8].

1.1 Phase Diagram: Understanding the Quantities of Components Involved in Ionotropic Gelation

In the formulation process involving polymers and active ingredients for particle formation in ionotropic gelation, it is crucial to carefully balance the concentrations of the polymer solution and the crosslinking ionic solution. To determine the optimal region for gel formation, it is recommended to construct a phase diagram based on the concentrations of the polymer and the crosslinking salt. Other factors such as pH and ionic strength can also influence the kinetics of crosslinking and the strength of the gel formed, as well as the porosity of the network [9].

The phase diagram defines three interaction zones between the polymer and the crosslinking ion: the "liquid" region, where the polymer and ion combinations fail to form a cohesive and firm structure; the "gel" region, where the reaction rate, structure, and water retention of the gel are optimal, indicating favorable polymer and ion combinations for particle formation; and the "syneresis" region, where a gel is formed, but water is expelled from the gelled matrix due to excessively high concentrations of crosslinker and polymer, resulting in a dense matrix that can contract and expel the solvent [9, 10].

1.2 Droplet Formation: The "Heart" of the Process

The initial phase of compound encapsulation via ionotropic gelation involves forming droplets composed of a gelling polymer and the active compound. These droplets come into contact with the crosslinking solution and solidify, creating the final particle structure. In ionotropic gelation, two types of droplet structures—and consequently particles—can be formed: spheres and true capsules. To create spheres, the active compound and polymer are mixed together and shaped into droplets that are then crosslinked, forming the final particles. For true capsules, the polymer solution must coat the active compound, and these structures can be produced using coextrusion systems. Figure 1 illustrates the concepts of simple extrusion (Fig. 1a) and coextrusion (Fig. 1b) for forming spheres and true capsules through ionotropic gelation.

Regarding the production of sphere-type particles, droplet formation can be achieved through simple extrusion using syringes or pipettes, utilizing nozzles with varying orifice sizes to obtain different particle diameters. Spraying with dual-nozzle device, emulsification, microfluidic devices, among others, can also be used for the formation of precursor droplets to form spheres. For the production of true capsules, droplets are formed within

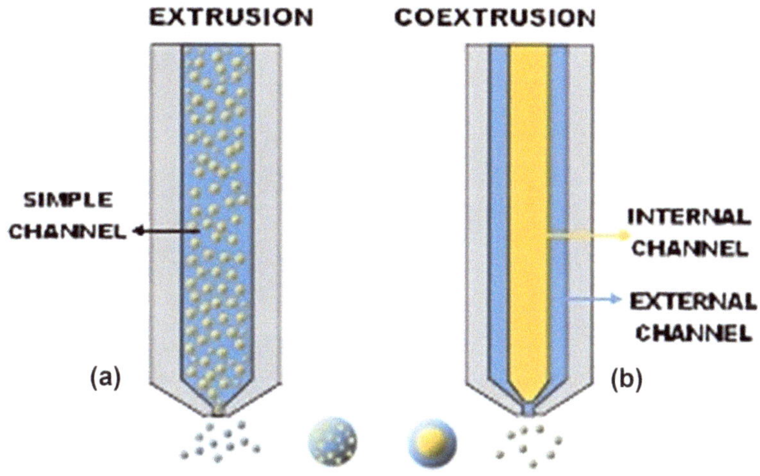

Fig. 1 Droplet formation by simple extrusion (**a** spheres) and coextrusion (**b** true capsules) in the production of particles through ionotropic gelation

concentric channels, where the polymer solution is pumped through an outer channel and the active material through an inner channel. Triple-fluid spray nozzles, concentric nozzles, and microfluidic devices are commonly used to create true capsule-type particles. Figure 2 shows the droplet system for the production of sphere-type and true capsule particles.

1.3 Crosslinking: Obtaining the Particles

Two techniques are used for the production of particles through ionotropic gelation: external gelation and internal gelation.

In external gelation, the polymer solution containing the active compound, in droplet form, comes into contact with the calcium ion solution. Instantly, an outer film of crosslinked polymer is formed, confining the remaining material and pre-structuring the particle. The calcium ions diffuse through this membrane and continue the crosslinking process until the particle is fully stabilized. The curing time required for particle stabilization depends on several factors, including the concentration of polymer and calcium and the droplet size. After curing, the particles are removed from salt solution and washed with water.

For internal gelation, a salt with a low degree of dissociation at the natural pH of the polymer solution is used as a source of calcium ions, which does not initially promote crosslinking. The solution containing the polymer, non-ionized salt, and active compound is emulsified in oil, forming microdroplets. An acidic solution is then introduced into the emulsion, migrating to the interfacial tension region between the oil and water. The acidic solution begins to lower the pH at the surface of the droplets, causing the salt to ionize and release calcium ions, which then crosslink the polymer and stabilize the particle. Once crosslinking

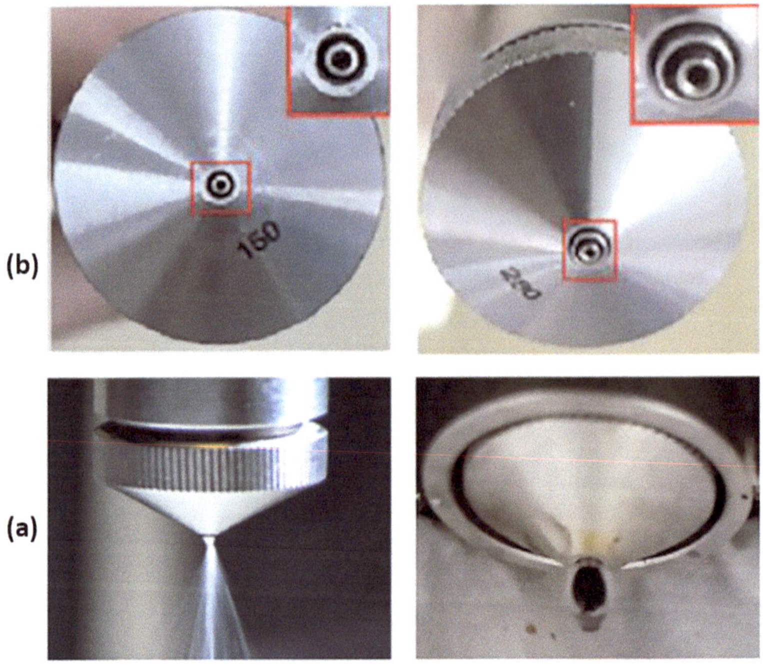

Fig. 2 Droplet formation systems for the production of sphere-type and true capsule-type particles. (**a**) Spheres—simple channel (double-fluid nozzle—atomization) and (**b**) capsules—concentric channels (triple-fluid nozzle without compressed air—dripping configuration)

is complete, the stabilized particles are removed from the oil, rinsed, and stored. Figure 3 shows the crosslinking of particles through external and internal ionotropic gelation.

This protocol aims to introduce the technique of microencapsulation through ionotropic gelation using alginate or pectin. It includes a discussion of key theoretical concepts to enhance understanding of the method. The practical steps for executing this technique will be outlined to ensure accurate reproduction and implementation.

2 Materials and Reagents

In the encapsulation of bioactive compounds for food applications using ionotropic gelation, sodium alginate and pectin are among the most commonly used polymers. These anionic polymers react with divalent cationic ions to form gels. This protocol outlines the procedures for particle formation using these polymers, and therefore, some general characteristics of these materials are described below.

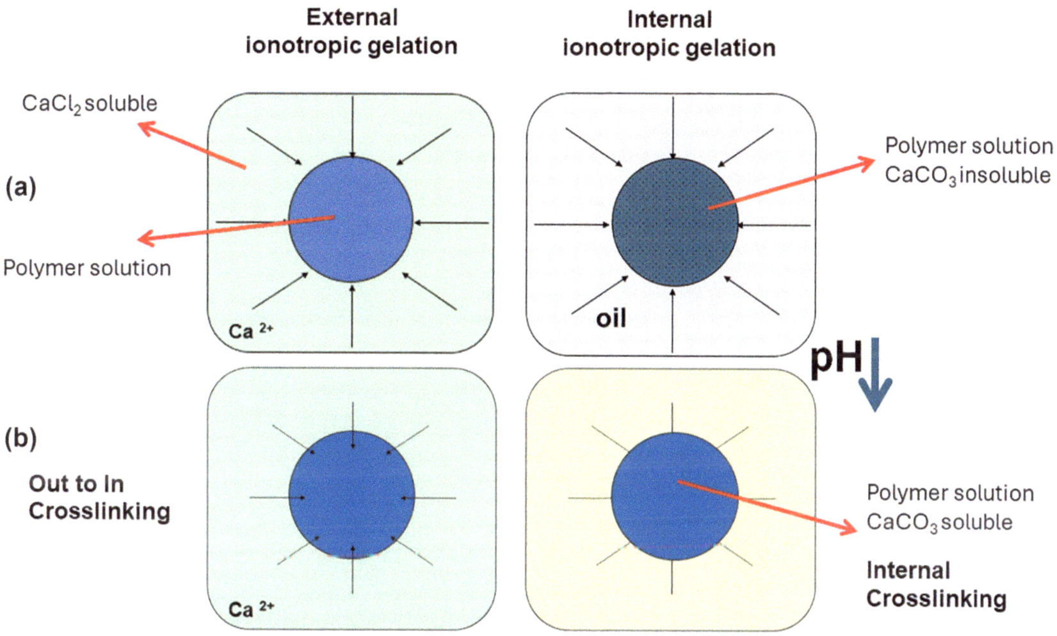

Fig. 3 Hardness of droplet of polymer in particles by crosslinking process: (**a**) external ionotropic gelation and (**b**) internal ionotropic gelation

2.1 Alginate

Sodium alginate (a polyanion) is the most commonly used encapsulating material in this technique. When it comes into contact with calcium (a divalent cation), it forms a membrane that is slightly soluble in water but permeable [10, 11].

Alginate is originally a component of the cell walls of brown algae, where it provides structural support by imparting rigidity while maintaining flexibility. It can constitute up to approximately 40% of the algae's composition. Alginate acid, the commercially extracted component from algae, consists of linear chains of β-D-mannuronic acid (M) residues linked by (1 → 4) bonds and α-L-guluronic acid (G) residues. The proportion of these residues can vary depending on the source of the algae, influenced by factors such as temperature, water salinity, and sunlight exposure [12, 13]. Alginate acid has low stability, and to produce a stable commercial product, it is combined with salts to form various types of alginates, such as sodium alginate (alginate acid + Na_2CO_3), potassium alginate (alginate acid + K_2CO_3), and calcium alginate (alginate acid + Ca_2CO_3) [12].

The composition of alginate classifies it as an anionic polymer that is water-soluble and capable of interacting with divalent cationic ions (especially Ca^{+2}) to form thermoresistant gels.

In addition to alginate derived from algae, biotechnologically produced alginate using bacteria has also been studied. These studies indicate that although productivity is still low and the material exhibits a higher degree of acetylation, the improved control over production conditions and greater batch standardization make this alginate source an attractive option from an industrial perspective [13, 14].

Alginate is widely used across various industrial sectors due to its structuring, gelling, viscoelastic, and film-forming properties. In the food industry, it serves as a gel former, thickener, water-binding agent, stabilizer, and film former, and is found in products such as desserts, tomato sauces, ice creams, mayonnaises, salad dressings, juices, meat coatings, and biscuits [12].

In ionotropic gelation, alginate forms droplets that complex with calcium ions, creating stable spherical structures capable of retaining hydrophobic substances or those with low water solubility. However, due to the gel's porosity, this system is not ideal for retaining low molecular weight hydrophilic substances. For that, strategies such as double emulsion, internal gelation, and reducing porosity through polymeric coating of the gel surface can be employed to control the release of these components.

2.2 Pectin

Pectin constitutes a group of substances with significant interest to the food industry. In recent decades, these compounds have been primarily used in powdered form as a valuable ingredient due to their ability to act as gelling agents. As a constituent of all terrestrial plants, pectin has been part of the human diet since the dawn of humankind. It has been evaluated and declared harmless from a toxicological standpoint by the JECFA (Joint FAO/WHO Expert Committee on Food Additives). No ADI (acceptable daily intake) has been established for pectin, thus there are no limitations on its use from a toxicological perspective [15].

Pectin is a polysaccharide of plant origin composed of galacturonic acid linked in α ($1 \rightarrow 4$) bonds, with a variable degree of methyl-esterified carboxyl groups. Commercial powdered pectin can be classified as high methoxyl (HM), with more than 50% esterified groups in the chain (degree of esterification or DE), or low methoxyl (LM), with a DE of less than 50%. The degree of amidation indicates the percentage of carboxyl groups in the form of amides. The degrees of methoxylation and amidation strongly influence the functional properties, such as solubility, gelling capacity, and the temperature and conditions for pectin gelation [15].

Low methoxyl pectins form gels only in the presence of divalent cations without the need for added sugars [16, 17], therefore it is more suitable for ionotropic gelation. Treating pectin with ammonium dissolved in methanol converts some methoxyl groups into carboxyl groups, leading to the formation of low methoxyl amidated pectins. Amidated pectins can have 15% to 25% of carboxyl groups in the form of carboxamides [15]. De-esterified low methoxyl pectins (LM) require a relatively high concentration of calcium within very narrow limits to achieve a gel with optimal consistency, whereas amidated LM pectins offer greater flexibility in this regard.

For both types of pectins, increasing the calcium concentration can enhance gel strength and raise the gelling temperature, up to the point of pre-gelatinization, where the gelling temperature approaches the boiling point [15].

Pectin is a wall material widely used in ionotropic gelation due to its electrostatic characteristics and good ability to form gels in different chemical variations, providing excellent properties for use as a wall material. Additionally, it is a plant-based product, appealing to the growing vegan audience. Using pectin to encapsulate highly water-soluble active ingredients can result in premature release due to diffusion through the porous surface of the ionic gel particles. This issue can be mitigated by incorporating an additional hydrophobic protective agent, such as vegetable oils, providing double protection for the hydrophilic active ingredient [6].

2.3 Crosslinking Agents

As previously discussed, both pectin and alginate form gels with divalent positive ions, with Ca^{+2} being the most effective. While $CaCl_2$ is the most commonly used source of ionic calcium, it imparts a bitter taste, and its residues in the particles can negatively affect the sensory profile of ingredients containing this salt. A viable alternative is calcium lactate, which has a more acidic taste and may reduce the sensory impact.

Beyond the concentration of the crosslinking solution, the solubility of the salt used is a crucial factor in achieving proper dissociation during external ionotropic gelation, with pH adjustment of the crosslinking solution as needed. In contrast, during internal gelation, the salt mixed with the polymer solution should remain undissociated, and pH adjustments can also be made if necessary.

2.4 Particle Formation: Droplet Formation

2.4.1 Simple Droppers (Pipettes and Syringes)

A simple way to conduct preliminary ionotropic gelation tests is by using pipettes or syringes to form the precursor droplets for the particles. These are low-cost and easily accessible devices. The diameter of the pipette tip or syringe needle determines the droplet size, allowing for the production of samples with different particle sizes. The disadvantage of using pipettes and syringes for particle formation at a laboratory scale is the lack of process standardization and the difficulty in producing larger quantities of particles, as these devices are operated manually.

2.4.2 Dripper Nozzles (Simple Extrusion and Coextrusion)

Among the ionotropic gelation techniques available, particle production through nozzles offers the advantage of adjustable parameters, allowing for the creation of particles with more controlled sizes and enabling high production rates of uniformly sized particles [8, 18, 19]. Furthermore, the use of specialized nozzles that pump the emulsion, dispersion, or solution in a laminar jet allows for precise and simultaneous control over the size of the microparticles [19].

Dripper nozzles can be used for simple extrusion, forming spheres, or for coextrusion, forming true capsules. Coextrusion involves the simultaneous extrusion of multiple substances with complementary characteristics to produce a final product. In microencapsulation, the most common application of this technique involves extruding the active compound through an inner channel while the polymeric wall material is extruded through an outer channel. This process generates particles with a core-shell structure, commonly referred to as true capsules in encapsulation.

2.4.3 Spray Nozzles (Dual-Fluid or Triple-Fluid Nozzles)

In spray nozzles, compressed air interacts with the liquid as both are extruded through an orifice, generating microdroplets in a spray cloud. This type of nozzle allows for the production of much smaller particles compared to dripping methods. The pressure of the compressed air can be adjusted to control the range of droplet sizes, and consequently, the range of particle sizes produced. Higher pressures typically result in smaller particles, while lower pressures produce larger ones. The viscosity of the polymer solutions also affects spraying performance—highly viscous solutions may produce larger particles and, in some cases, may not be adequately sprayed.

2.4.4 Feeding Devices

To supply systems that generate precursor droplets for particles, gravity feeding can be employed, where a container holding the polymer solution is attached above the dripping or spraying apparatus, allowing the liquid to flow downwards by gravity. This method is highly dependent on the liquid's density and viscosity. While gravity feeding offers a simple and low-cost solution, it can result in uneven feeding. A more effective alternative for handling more viscous fluids is to use containers under positive pressure to feed the dripper or sprayer nozzles, ensuring a more consistent flow.

Systems such as peristaltic pumps are widely used for feeding dripper/sprayer nozzles. These devices allow for good flow control with continuous feeding; however, they produce a kind of "pulse" in the liquid flow, which can affect the polydispersity of the particles produced.

In contrast, syringe pumps provide very precise feeding with a continuous and consistent flow, but they are limited by the volume of the syringes, which restricts large-scale production.

3 Conducting the Experiments

3.1 Preparation of Polymer Solutions

Emulsions (Water in Oil)

The ionotropic gelation technique, when preceded by emulsification using vegetable oil, effectively creates a barrier against the loss of hydrophilic compounds [5].

An emulsion consists of two immiscible liquids (typically water and oil), where one liquid is dispersed as small droplets within the other, with droplet diameters typically ranging from 100 nm to 100 μm [20]. Generally, multiple W/O/W emulsions are prepared in two stages: first, a simple W/O emulsion is created using a homogenization process and an emulsifier with a low hydrophilic-lipophilic balance (HLB) value to stabilize the emulsion. In the second stage, the initial emulsion is homogenized with another aqueous phase using a hydrophilic emulsifier to stabilize the double emulsion (W1/O/W2) [21–23].

Maintaining the stability of both the simple W/O emulsion and the double W/O/W emulsion is a key challenge in applying ionotropic gelation [8]. Double or multiple emulsions offer several advantages over simple emulsions, including the ability to produce foods with sensory and physicochemical properties similar to those of simple emulsions, while also serving as delivery systems for controlled release of active agents such as flavors, antimicrobial agents, or bioactive lipids [24].

To achieve emulsions with differentiated and desirable structures, it is crucial to first understand and monitor the interactions between ingredients and the impact of the process on the rheology and microstructure of the final product, as well as during storage. High-pressure homogenization promotes the formation of emulsions with very small droplet sizes, making it an effective method for preparing multiple emulsions [24]. To form stable emulsions, emulsifiers—chemical species (or mixtures) that adsorb onto the surface of droplets during homogenization—are used to create protective layers that prevent aggregation of the dispersed phase and reduce interfacial tension [25].

3.2 Preparing the Crosslinking Solutions

Calcium is the most commonly used cation in crosslinking solutions due to its lower toxicity when compared to other metals. The use of $CaCl_2$ at a concentration of 2–3% is generally sufficient to ensure crosslinking. To improve crosslinking and particle stability, the crosslinking solution is often acidified to a pH close to 3.0. The solution should be stirred, and the particles should remain immersed in the solution for 10 to 15 min to ensure that all particles are properly crosslinked.

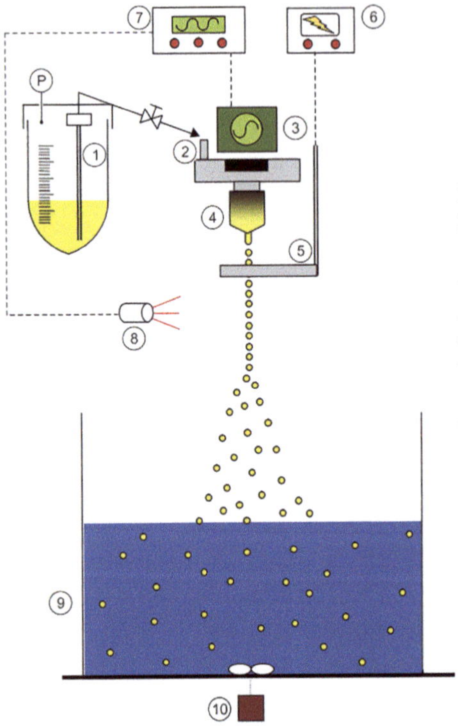

Fig. 4 Ionotropic gelation by dripping—Büchi B-390 Encapsulator. (*1*) Mixing the emulsion containing the active ingredient with the pectin solution (polymer), (*2*) feeding the emulsion with the aid of compressed air, (*3*) application of a superimposed vibration on a laminar film of the emulsion, (*4*) droplet formation, (*5*) electrostatic alteration causing droplet dispersion, (*6*) dispersion control, (*7*) vibration control, (*8*) visualization of the droplet formation under stroboscopic light, (*9*) microparticle formation in the crosslinking solution, (*10*) magnetic stirrer , (*11*) air pressure

3.3 Particle Production at Laboratory Scale

External gelation

For the production of particles by external ionotropic gelation using the dripping technique, an Encapsulator B 390 (Büchi, Essen, Germany) can be used (Fig. 4). This device operates on the principle of a fluid jet with laminar flow, which is subjected to a specific frequency, breaking it into uniform droplets. These droplets are then solidified through chemical or physical processes.

The nozzle size for this method can range from 80 to 1000 μm, with the choice depending on the viscosity of the sample or emulsion to prevent clogging. Key adjustable parameters include vibration frequency (40–6000 Hz), electrode voltage (250–2500 V), and a fixed feed pressure, all aimed at producing particles with desirable morphological and color characteristics. The distance between the nozzle and the surface of the calcium solution was set at 10 cm, and the curing time for the particles in the calcium solution was 15 min, under agitation.

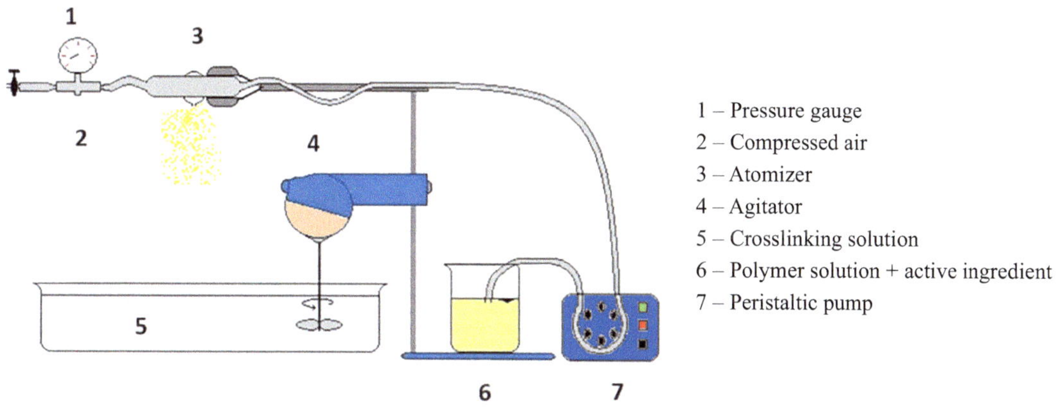

Fig. 5 Ionotropic gelation using the spray dryer atomizer. (*1*) Pressure gauge, (*2*) compressed air, (*3*) atomizer, (*4*) agitator, (*5*) crosslinking solution, (*6*) polymer solution + active ingredient, (*7*) peristaltic pump

For the production of particles by external ionotropic gelation using the spraying technique, a dual-fluid atomizer (mini Spray Dryer B-290, Büchi, Essen, Germany) with a diameter of 0.7 mm can be used, installed outside the spray chamber (Fig. 5). The distance between the nozzle and the surface of the calcium solution is set at 18 cm, and the curing time of the particles in the calcium solution is 15 min. The operating temperature is ambient. The air pressure can be adjusted between 20 and 30 mm (0.15–0.23 bar) and the feed rate between 20% and 50% (0.70–1.61 mL/min), with the goal of obtaining particles with good morphological characteristics and color.

4 Main Characterization Analyses for Particles Obtained by Ionotropic Gelation

After producing particles via ionotropic gelation, several analyses should be conducted to confirm that the encapsulation objectives have been met. These analyses can be categorized into two main groups: those related to the particle characteristics and those related to the application and function of the particles.

For particle characteristics, encapsulation efficiency can be used to determine the amount of active compound retained after production. Morphological analyses, size descriptors, and particle size distribution can be employed to evaluate the appearance and structure of the particles. Additional characteristics, such as moisture content and water activity, may also be assessed depending on the type of particles.

Further analyses can be conducted depending on the intended application of the particles. Properties such as color, texture, digestibility, active compound release rates, and the protection of the active compound can be studied to evaluate particle performance in specific applications. For particles produced using ionotropic gelation, the following analyses are recommended for thorough characterization:

- *Moisture (%):* The moisture content of the particles can be determined using the gravimetric method under heating until reaching a constant weight. Precautions during the procedure: When weighing the sample, it is crucial to remove any external water remaining from the crosslinking or rinsing process. Use absorbent paper to quickly blot away the excess water, taking care not to compress the particles to avoid removing the water retained inside them. The sample should be evenly spread on the plate before being placed in the oven. Particle agglomeration can impede water evaporation, potentially compromising the accuracy of the analysis.
- *Encapsulation efficiency (%):* Encapsulation efficiency in ionotropic gelation measures the correlation between the amount of active compound present in the particles and the amount initially added to the formulation. A critical requirement in this analysis is the efficient disruption of the particles to fully release the active compound for accurate recovery and quantification. Disruption can be achieved using calcium ion sequestrants, such as citrate salts or EDTA, or through mechanical means, such as cell disruptors. In all cases, it is essential to ensure that the disruption method does not alter the active compound.
- *Particle size distribution:* Understanding particle size and its distribution is essential for evaluating both applicability and performance. Particle size influences the release mechanisms of the active compound and affects sensory perception, both visually and in terms of mouthfeel. In ionotropic gelation, the size range achieved through dripping or spraying can be assessed by determining the average diameter and size distribution using techniques such as optical microscopy or laser diffraction. Sieve-based techniques are not recommended, as the moisture in the particles can cause them to adhere to the sieve meshes, leading to distorted results. Moreover, the water on the outer surface of the samples can further interfere with the accuracy of the measurements.
- *Particle morphology:* The appearance of particles produced by ionotropic gelation is a key characteristic. Techniques such as optical microscopy, confocal microscopy, and scanning electron microscopy enable detailed observation of the active compound's distribution within the particle structure, the potential release of the compound through the membrane, and the overall integrity of the particle. These microscopy techniques are valuable for observing particles within food matrices during application, providing insights into their behavior in real-world

conditions. Microscopy can also serve as a complementary tool in digestibility tests, revealing structural changes in the particles during simulated digestion.

- *FT-IR spectroscopy:* FT-IR (Fourier transform infrared) spectroscopy is a highly sensitive and high-resolution method of infrared spectroscopy. This technique relies on the interference of radiation between two beams, producing an interferogram. An interferogram records the signal generated by the combination of multiple possible frequencies, which are then processed mathematically using the Fourier transform to convert the data into a spectrum. The primary purpose of obtaining FT-IR absorption spectra is to measure a sample's ability to absorb radiation across a wide range of wavelengths.

FT-IR is used to characterize individual components of microparticles and to assess potential molecular interactions between the contents of the carriers. The microparticle spectrum in this technique is divided into three regions: (i) the protein region (located between 1700 and 1480 cm^{-1}), which corresponds to C=O, C–N, and N–H bonds; (ii) the carbohydrate region (located between 1200 and 870 cm^{-1}), which refers to the absorption bands of C–O, C–H, and C–O–H; and (iii) the region between 3500 and 2850 cm^{-1}. Peaks between 1400 and 1600 cm^{-1} are associated with aromatic carbon double bonds and C=C stretching or are attributed to the C=O stretching vibration of flavonoids [26].

Samples should be pre-dried in a vacuum oven at 70 °C for 24 h. For particles, extracts, and pectin, the commonly employed method is potassium bromide (KBr) pelletization (100 mg of KBr to 1 mg of sample). For oils, the generally used method is a liquid film (KBr window with 1 drop of sample). FT-IR spectra are collected in transmittance (T%) across the wavelength range of 400 to 4000 cm^{-1}, with a resolution of 4 cm^{-1}.

5 Main Challenges

During the tests for obtaining particles by ionotropic gelation, some issues may arise, limiting or even preventing the formation of the particles. In this section, we will discuss some of these problems.

5.1 Obstruction in Dripping or Spraying

The presence of solid particles in suspension within the active compound or polymer solution can lead to obstructions in dripper or sprayer nozzles, directly impacting process productivity and potentially damaging equipment.

To prevent such issues, a particle size analysis can be conducted to determine the appropriate nozzle size for particle production. Ideally, if nozzle variation is possible, select a nozzle with a diameter at least twice the size of the largest particle fraction, using the 90th percentile diameter (D90) as a reference. If changing the nozzle size is not feasible, the size of suspended particles can be reduced using colloidal mills, Ultra-Turrax, or even a blender to break down the particles and minimize clogging.

Sieving can also be employed to separate larger particles, although this method may inadvertently remove components essential to the material's function or activity.

5.2 Retention of Hydrophilic Substances

Gelation presents challenges for encapsulating active compounds with high water solubility, such as phenolic compounds, due to the risk of premature diffusion through the porous microsphere [27]. However, these challenges can be mitigated by employing a hydrophobic barrier, such as multiple emulsions. This microencapsulation technique is used for both hydrophilic and lipophilic active compounds and functional foods, enhancing the protection and retention of the desired compounds [23, 28–30].

Various strategies have been proposed in the literature to address the low retention of microparticles obtained through ionotropic gelation. Incorporating fillers like starch, inulin, or other polysaccharides has been shown to improve the retention, as well as the mechanical, morphological, and physical properties of microparticles ranging from 2 to 4 mm [31, 32]. Silveira et al. [33] studied pectin-starch spheres produced by the subsequent adsorption of pomegranate extract. The authors found that reducing the concentration gradient between the internal and external environments significantly improved retention to approximately 101%, resulting in greater microstructural stability and slower release rates. They suggested that the combined use of adsorption and coating techniques in starch-containing spheres could be a promising approach to enhancing the retention of hydrophilic compounds. However, they also noted that certain adjustments in these techniques are essential for the practical application of these polyphenol-loaded spheres in food matrices.

5.3 Hydrocolloid Properties for Different Batches, Types, or Brands

The lack of standardization in commercially available biopolymers has been noted in numerous studies. Both alginate and pectin are significantly influenced by the source material used in their production, which can impact their crosslinking properties. For instance, in the case of alginate, variations in climatic conditions from year to year can affect its viscosity, composition, and distribution of reactive groups, potentially altering the characteristics of particles produced from different batches of polymers.

In short-term studies, it is advisable to acquire a sufficient quantity of materials to complete the entire study using the same batch. When this is not feasible, initial characterizations of the materials should be conducted to allow for comparisons with new batches and to anticipate any potential differences. Viscosity and gel strength analyses are particularly useful as reference points when comparing polymers from different batches.

5.4 Crosslinking Interferences

Some substances in the material to be encapsulated may react with the wall polymer, leading to premature crosslinking and making encapsulation unfeasible. A practical example observed by our group, though not yet published, involved the encapsulation of fish feed via ionotropic gelation. The feed, which contained mineral salts, prematurely induced pectin crosslinking during preliminary tests, impairing an adequate particle production. To overcome the problem, a pre-solution of feed and pectin was made to neutralize the salt present in the feed, which was mixed with the pectin used as the wall material for particle production.

6 Drying Possibilities: Fluidized Bed, Freeze-Drying, and Oven-Drying

Partial post-drying can be a strategy to increase the microbiological stability of microparticles obtained through ionotropic gelation. Particles with lower moisture content can be used in drier products, preventing changes due to high moisture content, along with the possibility of using a larger number of particles, thereby bringing greater potential for bioactive compounds to food products [34].

The advantages of fluidized bed drying over other types of processes include increased drying rates, thermal efficiency of the process, relatively low cost, and greater ability to control the product's temperature, allowing the use of milder temperatures [35]. However, this process has some challenges, such as the effect of particle size on the drying endpoint, leading to uneven drying and inaccurate determination of the drying endpoint [23].

Currently, this type of dryer is also used for coating microparticles in various areas of application, which can enhance the protection of the microencapsulated active compound, characterizing this process as a highly thermally efficient and easy-to-operate technology [36–38].

Freeze-drying is effective in preserving the spherical structure of the particles with minimal alterations. The initial freezing step should be performed quickly to avoid damaging the polymer matrix, as slow freezing and the resulting formation of large ice crystals can damage the gel, similar to what occurs in cells. Depending on the sample size, immersion in liquid nitrogen may suffice for rapid freezing before freeze-drying. Ultra-freezers with rapid freezing capabilities can also be used.

For small sample quantities, drying in a laboratory-scale oven is feasible as long as gentle drying conditions are maintained (low temperatures, below 50 °C) with adequate air circulation. Rapid water evaporation at high temperatures can damage the particle structure, causing them to flatten or the gel to break down. While alginates are thermally resistant and can better withstand oven-drying, milder temperatures are still recommended. It's important to consider that oven-drying may also affect the active compound.

Both freeze-drying and oven-drying can lead to particle adhesion, resulting in damage such as breakage or the formation of clumps (structures where multiple particles stick together) during separation after drying. This adhesion or breakage can interfere with the determination of size distribution, morphology, active compound release, active compound protection, and other properties that depend on the real and individual structure of the particles. The use of small amounts of anti-caking agents, such as glycerol, in the wet sample can help prevent adhesion.

7 Examples of Particle Production and Their Applications

7.1 Hibiscus Extract: Applications in Yogurt and Jelly Candy

The application of hibiscus extract microparticles obtained by ionotropic gelation using dripping and spraying techniques [8] was conducted in jelly candy [39] and natural yogurt [40].

In the jelly candy study, [39] prepared two batches of jellies containing 7.14 g/100 g of anthocyanin microparticles obtained by atomization and 7.14 g/100 g of anthocyanin microparticles obtained by dripping. Figure 6 shows the final appearance of the jelly candies with hibiscus extract microparticles.

The application of microparticles in the jelly candy matrix demonstrated technical viability, providing color and functionality, with an average retention of total anthocyanins of 56% in the candy matrix. Both types of microparticles showed good thermal resistance within the matrix and remained intact throughout the process.

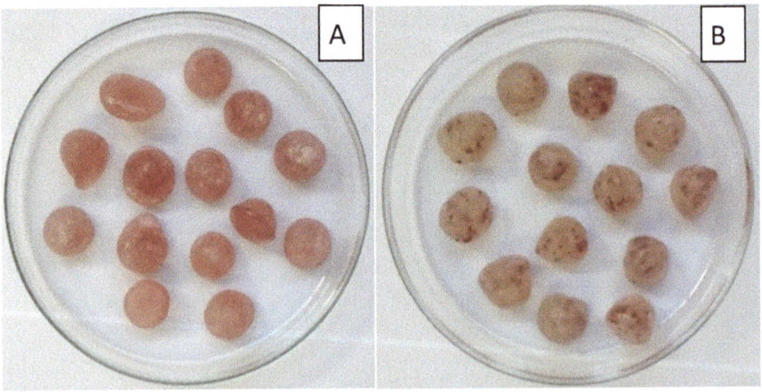

Fig. 6 Jelly candies with the application of microencapsulated hibiscus extract, obtained by atomization (**a**) and dripping (**b**)

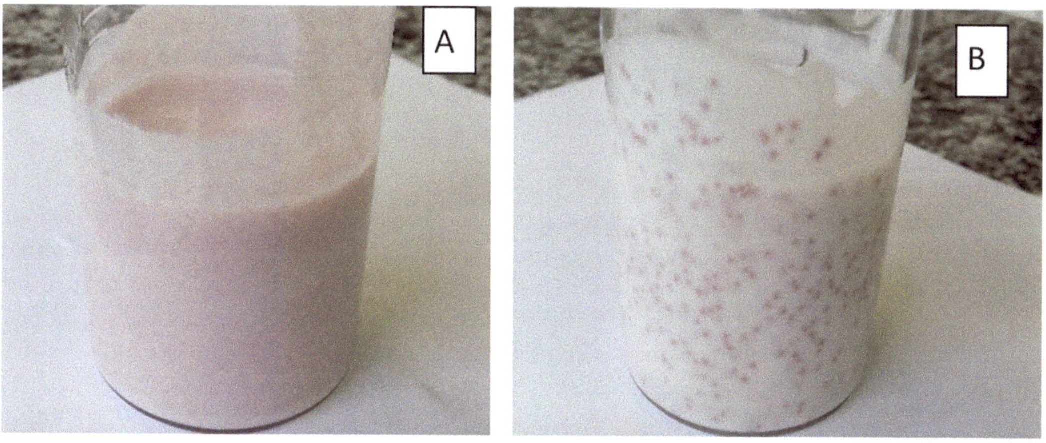

Fig. 7 Yogurts with the application of microencapsulated hibiscus extract, obtained by atomization (**a**) and by dripping (**b**)

Jelly matrices with particles generated by the atomization technique had higher acceptance of the attribute appearance, with an average score of 6.2, corresponding to values between "I liked it" and "liked it slightly."

During the 62-day shelf life, the anthocyanin microparticles provided color to the jelly matrix, with retention of 23–73% of anthocyanins, with higher retention for jellies with particles generated by dripping.

The jelly candy matrix with the addition of particles generated by ionotropic gelation using the dripping technique showed higher bioactive retention, although it presented a less homogeneous color distribution.

In the yogurt study [40], two yogurt formulations were prepared, containing 20 g/100 g of hibiscus extract microparticles obtained by atomization and 20 g/100 g of hibiscus extract microparticles obtained by dripping. Figure 7 shows the final appearance of the yogurts containing hibiscus extract microparticles.

The application of microparticles in the yogurt matrix demonstrated technical viability, providing color and functionality, with an average retention of total anthocyanins of 60% in the yogurt matrix. The addition of microparticles did not negatively affect the viability of lactic bacteria in the matrix, favoring their maintenance.

The yogurt matrix with particles generated by atomization showed higher acceptance of the attribute appearance, with an average score of 6.5, corresponding to values between "liked it" and "liked it slightly."

During the 62-day shelf life, the anthocyanin microparticles provided color to the yogurt matrix, with retention of 48–21% of anthocyanins, proving the efficacy of the encapsulated hibiscus extract as a coloring agent.

The yogurt matrix with the addition of particles generated by ionotropic gelation using the dripping technique showed higher retention of bioactive compounds, although they presented less homogeneous color distribution.

7.2 Green Tea Extract: Application in Ice Cream

Cutrim et al. [41] and Cutrim [42] studied the double emulsion technique reported by Moura et al. [8] for the encapsulation of green tea extract by ionotropic gelation for application in ice cream. The microparticles were produced with pectin at a concentration of 3 g/100 g, and the active compound was green tea powder reconstituted in a 1:1 ratio with filtered water, emulsified in canola oil at a 30:70 ratio, respectively, using 4 g/100 g of PGPR as a surfactant. This initial emulsion was added to the pectin in a 20:80 ratio, followed by spraying through a dual-fluid nozzle into a calcium chloride solution at 3 g/100 g, acidified to pH 3 with citric acid. The spraying conditions were: atomization pressure of 0.15 bar, nozzle feed rate of 1.3 mL/min, and a distance of 18 cm from the surface of the crosslinking solution. After 15 min of curing, the particles were collected by filtration using filter paper. The encapsulation efficiency, considering the total phenolic content, was 72.6%, and the appearance of the emulsions and particles can be seen in Fig. 8.

Relevant results from the application of ionotropic gelation particles in ice cream include an approximate 82% increase in total polyphenol content compared to the control ice cream without microparticles. In the initial sensory evaluation, the ice cream containing ionotropic gelation microparticles had an acceptance rate of 56.7%, with panelists noting the perception of particles during tasting. However, in a follow-up sensory session where participants were informed about the presence of green tea, the acceptance rate increased to 67.2%. This suggests that educating consumers about the benefits of encapsulated substances can enhance product acceptance, even if the particles introduce some sensory perceptions, such as grittiness. Further details of this study are reported in Cutrim (2019; http://app.uff.br/riuff/handle/1/28484).

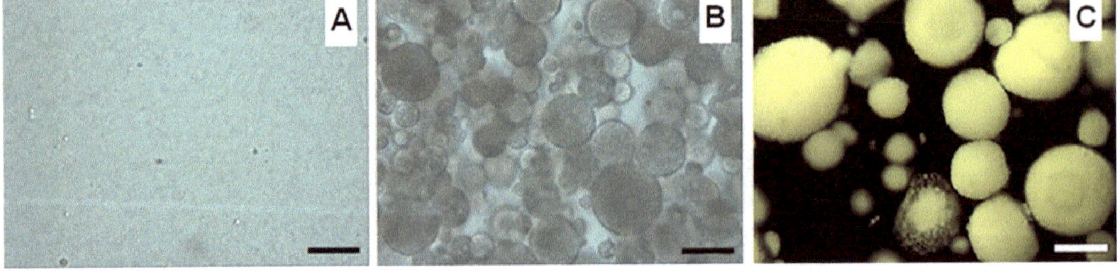

Fig. 8 Particles containing green tea obtained by ionotropic gelation using the double emulsion technique. (**a**) W/O Emulsion: W: reconstituted green tea; O: canola oil. (**b**) W/O/W double emulsion: W: reconstituted green tea; O: canola oil; W: pectin 3 g/100 g. (**c**): ionotropic gelation particles. Images (**a**) and (**b**): 1000× magnification and 20 μm scale bar. Image (**c**): 100× magnification and 200 μm scale bar

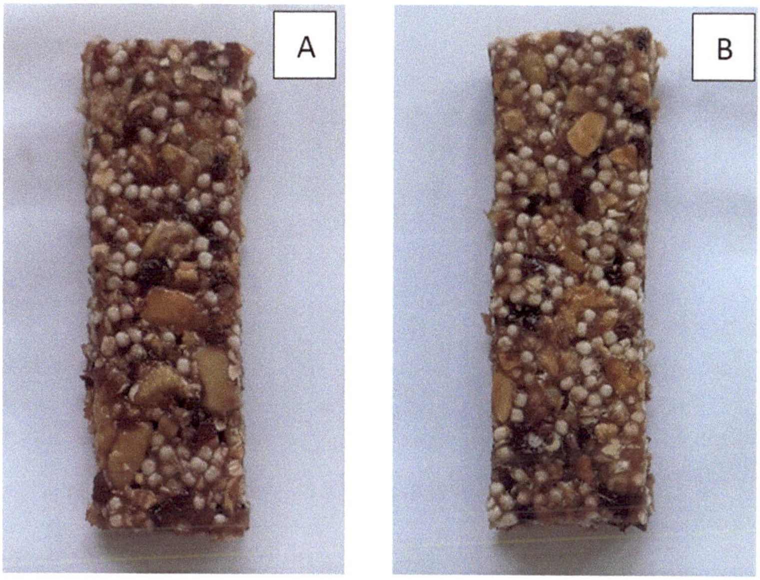

Fig. 9 Fruit-filled cereal bars with the addition of wet (**a**) and dry (**b**) yerba mate particles

7.3 Mate Extract: Application in Fruit Bars and Sausages

Microparticles obtained by ionotropic gelation using dripping techniques and subsequent drying in a fluidized bed [34] were incorporated in fruit-filled cereal bars [43] and Vienna sausages [44].

In the study with fruit-filled cereal bars [43], two different formulations were made to maintain the proportion of phenolic compounds: bars with the addition of 9% wet mate particles and bars with the addition of 4% dry mate particles. Figure 9 shows the final appearance of the fruit-filled cereal bars with mate microparticles.

The fruit-filled cereal bar with the addition of extract and dry particles had an acceptance rate above 80%, while the bar made with wet particles was the least accepted, with a 25% rejection rate. The drying of the microparticles was crucial because, regarding texture—a key parameter for this type of product—the bar made with wet particles was rejected by 46.9% of the assessors. Additionally, 84% of the assessors considered the sample to be less crunchy than ideal. Therefore, the use of combined techniques such as microencapsulation of the mate extract, complemented by fluidized bed drying, contributed to providing products with better-preserved properties of the compounds of interest and a high acceptance rate among consumers.

Regarding the Vienna sausages, Wensing et al. [44] produced sausages with the addition of 0.1% free yerba mate extract and 1% dry yerba mate particles. Figure 10 shows the appearance of the sausages made with dry mate microparticles and free extract.

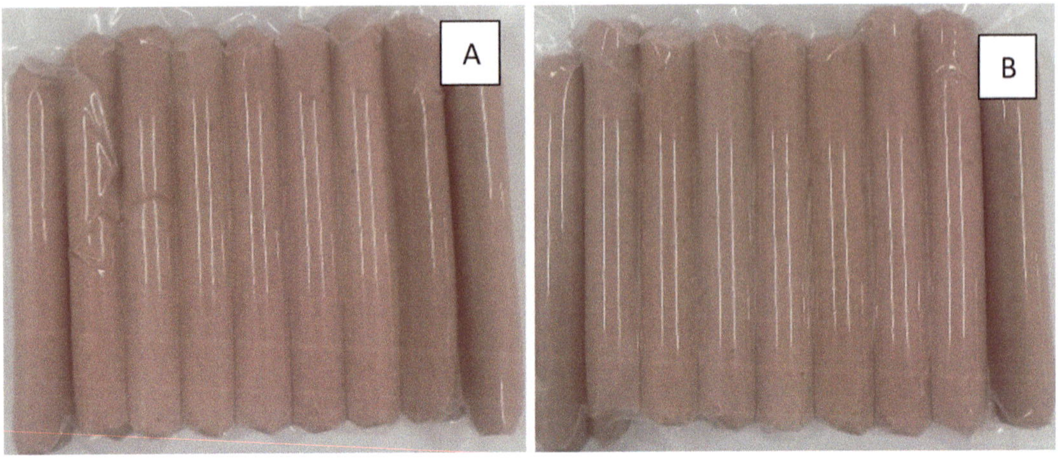

Fig. 10 Vienna sausages with free (**a**) and dry (**b**) yerba mate extract particles

The application of mate extract, both in its free and microencapsulated forms, has shown significant potential as a natural antioxidant and antimicrobial agent in Vienna sausages. The microencapsulated extract offered slightly greater benefits, effectively preserving antioxidant activity, phenolic compounds, color, and pH, while also inhibiting lipid oxidation. The microencapsulation technique allowed for the controlled release of compounds over time, providing extended protection against microbial growth.

Additionally, sensory evaluation using the cut-off point methodology revealed that all samples remained acceptable after 60 days of storage. However, the Vienna sausages containing microencapsulated particles (VP) showed a slight decline in average acceptability scores for "flavor" and "overall impression" at the final evaluation.

These findings underscore the promising potential of mate extract as a natural preservative for sausages. This approach offers substantial benefits in terms of antioxidant stability, antimicrobial properties, and the preservation of sensory qualities throughout the product's shelf life. These results contribute to the development of more natural and sustainable food preservation methods, aligning with the growing demand for healthier and safer products. Other examples include the application of yerba mate extract in yogurt [45].

References

1. Patil JS, Kamalapur MV, Marapur SC, Kadam DV (2010) Ionotropic gelation and polyelectrolyte complexation: the novel techniques to design hydrogel particulate sustained, modulates drug delivery system: a review. Dig J Nanomater Biostruct 5(1):242–248. https://www.researchgate.net/publication/265991547_Ionotropic_gelation_and_polyelectrolyte_complexation_The_novel_techniques_to_design_hydrogel_particulate_sustained_modulated_drug_delivery_system_A_review

2. Colak N, Torun H, Gruz J, Strnad M, Hermosín-Gutiérrez I, Hayirlioglu-Ayaz S, Ayaz FA (2016) Bog bilberry phenolics, antioxidant capacity and nutrient profile. Food Chem 201:339–349. https://doi.org/10.1016/j.foodchem.2016.01.062
3. de Moura SCSR (2018) Caracterização, perfil de liberação (in vitro) e aplicação de micropartículas contendo antocianinas do hibisco (Hibiscus sabdariffa L.) obtidas por gelificação iônica. Tese (Doutorado em Engenharia de Alimentos). Faculdade de Engenharia de Alimentos, Universidade Estadual de Campinas, Campinas. Disponível em: https://hdl.handle.net/20.500.12733/1634505. Acesso em: 29 ago. 2024
4. Mukai-Corrêa R, Prata AS, Alvim ID, Grosso CRF (2005) Caracterização de microcápsulas contendo caseína e gordura vegetal hidrogenada obtidas por gelificação iônica. Braz J Food Technol 8(1):73–80 https://www.yumpu.com/pt/document/view/14332953/caracterizacao-de-microcapsulas-contendo-caseina-e-ital
5. Kim ES, Lee JS, Lee HG (2016) Calcium-alginate microparticles for sustained release of catechin prepared via an emulsion gelation technique. Food Sci Biotechnol 25(5): 1337–1343. https://doi.org/10.1007/s10068-016-0210-8
6. Oehme A, Valotis A, Krammer G, Zimmermann I, Schreier P (2011) Preparation and characterization of shellac-coated anthocyanin pectin beads as dietary colonic delivery system. Mol Nutr Food Res 55:S75–S85. https://doi.org/10.1002/mnfr.201000467
7. Henning S, Leick S, Kott M, Rehage H, Suter D (2012) Sealing liquid-filled pectinate capsules with a shellac coating. J Microencapsul 29(2):147–155. https://doi.org/10.3109/02652048.2011.635220
8. Moura SCSR, Berling CL, Germer SPM, Alvim ID, Hubinger MD (2018) Encapsulating anthocyanins from *Hibiscus sabdariffa L.* calyces by ionic gelation: pigment stability during storage of microparticles. Food Chem 241: 317–327. https://doi.org/10.1016/j.foodchem.2017.08.095
9. Alvim ID, Prata AS, Grosso CRF (2016) Chapter 16: Methods of encapsulation. In: Garcia MPM, Gómez-Guillén MC, López-caballero ME, Barbosa-Cánovas GV (eds) Edible films and coatings: fundamentals and applications, 1st edn. CRC Press, Boca Raton, pp 299–315
10. Mukai-Corrêa R (2008) Produção de micropartículas por gelificação iônica para alimentação de larvas de peixe: estudos em sistema-modelo com inclusão de micropartículas lipídicas ou emulsão lipídica e testes *in vivo*. Tese (Doutorado em Alimentos e Nutrição). Faculdade de Engenharia de Alimentos, Universidade Estadual de Campinas, Campinas. Disponível em: https://hdl.handle.net/20.500.12733/1607152. Acesso em: 29 ago. 2024
11. Poncelet D (2006) Microencapsulation: fundamentals, methods and applications. Surf Chem Biomed Environ Sci 1:23–34. https://doi.org/10.1007/1-4020-4741-X_3
12. Onsoyen E (1997) Alginates. In: Thickening and gelling agents for food, 2nd edn. Ed. Alan Imeson, London, 320p
13. Garcia-Cruz CH, Foggetti U, Silva AN (2008) Alginato bacteriano: Aspectos tecnológicos, características e produção. Quim Nova 31: 1800–1806. https://doi.org/10.1590/S0100-40422008000700035
14. Urtuvia V, Maturana N, Acevedo F, Peña C, Díaz-Barrera A (2017) Bacterial alginate production: an overview of its biosynthesis and potential industrial production. World J Microbiol Biotechnol 33:198. https://doi.org/10.1007/s11274-017-2363-x
15. Food Ingredientes Brasil (2014) Pectinas: Propriedades e Aplicações, n. 29, 46–53
16. Be Miller JN, Huber KC (2008) Chapter 3 "Carbohydrates". In: Damodaran S, Parkin KL, Fennema OR (eds) Food chemistry, 4th edn. Marcel Dekker, New York, pp 84–151
17. Canteri MHG, Wosiack LMG, Scheer AP (2012) Pectina: da matéria-prima ao produto final. Polímeros Ciência e Tecnologia 22(2): 149–157. https://doi.org/10.1590/S0104-14282012005000024
18. Aizpurua-Olaizola O, Navarro P, Vallejo A, Olivares M, Etxebarria N, Usobiaga A (2016) Microencapsulation and storage stability of polyphenols from *Vitis vinifera* grape wastes. Food Chem 190:614–621. https://doi.org/10.1016/j.foodchem.2015.05.117
19. Menin A, Zanoni F, Vakarelova M, Chignola R, Donà G, Rizzi C, Mainente F, Zoccatelli G (2018) Effects of microencapsulation by ionic gelation on the oxidative stability of flaxseed oil. Food Chem 269:293–299. https://doi.org/10.1016/j.foodchem.2018.06.144
20. Mcclements DJ (2005) Food emulsions: principles, practices, and techniques, 2nd edn. CRC Press, Washington, DC
21. Akhtar M, Brent SM, Ehihumeme IA, Shiow HK (2014) Encapsulation of flavonoid in multiple emulsion using spinning disc reactor technology. Food Hydrocoll 34:62–67. https://doi.org/10.1016/j.foodhyd.2012.12.025

22. Pereira LJB, Garcia-Rojas EE (2014) Emulsões múltiplas: Formação e aplicação em microencapsulamento de componentes bioativos. Cienc Rural 45(1):155–162. https://doi.org/10.1590/0103-8478cr20140315
23. Santos AP (2024) Desenvolvimento de microesferas secas contendo ativos da pitanga (*Eugenia uniflora* L.) pela técnica de gelificação iônica. Dissertação (Mestrado em Engenharia de Alimentos), Faculdade de Engenharia de Alimentos, Universidade Estadual de Campinas, Campinas. Desenvolvimento de microesferas secas contendo ativos da pitanga (Eugenia uniflora L.) pela técnica de gelificação iônica | Unicamp
24. Perrechil FA (2008) Avaliação estrutural e reológica de emulsões simples e múltiplas estabilizadas por caseinato de sódio e jataí. Dissertação (Mestrado em Engenharia de Alimentos). Faculdade de Engenharia de Alimentos, Universidade Estadual de Campinas, Campinas, SP. Disponível em: https://hdl.handle.net/20.500.12733/1606424. Acesso em: 29 ago. 2024
25. Mcclements DJ, Decker EA, Weiss J (2007) Emulsion-based delivery systems for lipophilic bioactive components. J Food Sci 72(8):R109–R124. https://doi.org/10.1111/j.1750-3841.2007.00507.x
26. Van Hoogmoed CG, Busscher HJ, De Vos P (2003) Fourier transform infrared sprectroscopy studies of alginate-PLL capsules with varying compositions. J Biomed Mater Res 67:172–178. https://doi.org/10.1002/jbm.a.10086
27. Kurozawa LE, Hubinger MD (2017) Hydrophilic food compounds encapsulation by ionic gelation. Curr Opin Food Sci 15:50–55. https://doi.org/10.1016/j.cofs.2017.06.004
28. Beldarrain-Iznaga T, Villalobos-Carvajal R, Leiva-Veja J, Armesto ES (2020) Influence of multilayer microencapsulation on the viability of *Lactobacillus casei* using a combined double emulsion and ionic gelation approach. Food Bioprod Process 124:57–71. https://doi.org/10.1016/j.fbp.2020.08.009
29. Chen L, Gnanaraj C, Arulsevan P, El-Seedi H, Tend H (2019) A review on advanced microencapsulation technology to enhance bioavailability of phenolic compounds: based on its activity in the treatment of Type 2 diabetes. Trends Food Sci Technol 85:149–162. https://doi.org/10.1016/j.tifs.2018.11.026
30. Rakshit M, Srivastav PP (2022) Encapsulation of hydrolysable tannin from pomegranate peel in W/O/W double emulsion: in-vitro digestion, release kinetics, storage and physical stabilities. Food Hydrocoll Health 2:100067. https://doi.org/10.1016/j.fhfh.2022.100067
31. Córdoba AL, Deladino L, Martino M (2013) Effect of starch filler on calcium-alginate hydrogels loaded with yerba mate antioxidants. Carbohydr Polym 95(1):315–323. https://doi.org/10.1016/j.carbpol.2013.03.019
32. Busic A, Belscak-Cvitanovic A, Vojvodic Cebin A, Karlovic S, Kovac V, Spoljaric I, Mrsic G, Komes D (2018) Structuring new alginate network aimed for delivery of dandelion (Taraxacum officinale L.) polyphenols using ionic gelation and new filler materials. Food Res Int 111:244–255. https://doi.org/10.1016/j.foodres.2018.05.034
33. Silveira MP, Almeida FLC, Alvim ID, Silvia Prata AS (2023) Encapsulation of pomegranate polyphenols by ionic gelation: strategies for improved retention and controlled release. Food Res Int 174:113590. https://doi.org/10.1016/j.foodres.2023.113590
34. Budin AC, Takano LV, Alvim ID, Moura SCSR (2023) Stability of yerba mate extract, evaluation of its microencapsulation by ionic gelation and fluidized bed drying. Heliyon 9(6). https://doi.org/10.1016/j.heliyon.2023.e16611
35. Parikh DM, Drying. (2017) Em: how to optimize fluid bed processing technology. Elsevier, Cambridge, pp 55–63
36. Kim DH, Park WR, Kim JH, Cho EC, Na EJ, Kim J, Oh S (2012) Fabrication of pseudoceramide-based lipid microparticles for recovery of skin barrier function. Colloids Surf B: Biointerfaces 94:236–241. https://doi.org/10.1016/j.colsurfb.2012.01.049
37. Mcclements DJ (2014) Nanoparticle- and microparticle-based delivery systems, 1st edn. CRC Press, London
38. Paulo BB (2017) Estratégias para produção de partículas lipídicas em leito fluidizado. Dissertação [Mestrado em Engenharia de Alimentos]—Faculdade de Engenharia de Alimentos, Universidade Estadual de Campinas. Disponível em: https://hdl.handle.net/20.500.12733/1631122. Acesso em: 29 ago. 2024
39. Moura SCSR, Berling CL, Garcia AO, Queiroz MB, Alvim ID, Hubinger MD (2019) Release of anthocyanins from the hibiscus extract encapsulated by ionic gelation and application of microparticles in jelly candy. Food Res Int 121:542–552. https://doi.org/10.1016/j.foodres.2018.12.010
40. Moura SCSR, Schettini GN, Garcia AO, Gallina DA, Alvim ID, Hubinger MD (2019) Stability of hibiscus extract encapsulated by ionic

gelation incorporated in yogurt. Food Bioprocess Technol 12:1500–1515. https://doi.org/10.1007/s11947-019-02308-9

41. Cutrim CS, Alvim ID, Cortez MAS (2019) Microencapsulation of green tea polyphenols by ionic gelation and spray chilling methods. J Food Sci Technol 56:3561–3570. https://doi.org/10.1007/2Fs13197-019-03908-1

42. Cutrim CS Microencapsulação de Extrato de Chá Verde (Camellia Sinensis) para Aplicação de Polifenóis em Produtos Lácteos. Universidade Federal Fluminense. Tese de doutorado Avaliable in: http://app.uff.br/riuff/handle/1/28484

43. Budin AC (2022) Microencapsulação do extrato de erva-mate (Ilex paraguariensis) por gelificação iônica: avaliação da estabilidade e aplicação em barras de frutas e cereais. Dissertação de Mestrado em Ciência e Tecnologia de Alimentos. ITAL—Instituto de Tecnologia de Alimentos, Campinas. http://repositorio.ital.sp.gov.br/jspui/handle/123456789/607

44. Wensing C et al (2023) Avaliação da atividade antioxidante e antimicrobiana do extrato de erva-mate (ilex paraguariensis) microencapsulado, em salsicha. In: Anais do 15° SLACAN – Simpósio Latino Americano de Ciência de Alimentos e Nutrição, 2023, Campinas. Anais eletrônicos.... Galoá, Campinas. Disponível em: https://proceedings.science/slacan-2023/trabalhos/avaliacao-da-atividade-antioxidante-e-antimicrobiana-do-extrato-de-erva-mate-ile?lang=pt-br Acesso em: 29 Ago. 2024

45. Takano LV, Budin AC, Toti FGP, Alvim ID, Moura SCSR (2022) Chapter 7: Emprego da técnica de gelificação iônica na microencapsulação dos bioativos do extrato de erva-mate (ILEX PARAGUARIENSIS): aplicação das micropartículas em alimentos. In: Open science research IX. Editora Científica Digital, São Paulo, pp 109–123. https://doi.org/10.37885/221211384

Chapter 4

Encapsulation of Food Additives, Nutrients, Extracts, and Bioactive Compounds by Spray-Drying

Volnei Brito de Souza, Fabricio Luiz Tulini, and Carmen Silvia Favaro-Trindade

Abstract

Spray-drying is one of the oldest and most widely used dehydration techniques. It involves converting liquid feeds, such as solutions, dispersions, or emulsions, into powders. Additionally, spray-drying is a popular method for encapsulating or immobilizing compounds into polymeric matrices when a carrier is used. Among the carriers, maltodextrins, modified starches, and Arabic gum are the most commonly used materials. The particles produced by spray-drying are matrix-type, shrunken hollow spheres that vary in size from submicron to micron scale, depending on the atomizer employed. Using spray-dried particles as carriers for food compounds has several advantages, including protection against degradation, improved dispersibility or solubility, easier handling of labile liquids, reduced volatilization, and masking unpleasant tastes or odors. This protocol aims to guide the reader through the key steps of spray-drying encapsulation, including selecting carriers and operational conditions, providing an overview of the most critical influencing factors, and offering tips on how to overcome some of the process challenges.

Key words Atomizer, Drying process, Microparticles, Functional ingredients

1 Introduction

Spray-drying is a well-established and popular technology used to encapsulate or immobilize compounds in food using polymeric matrices. This continuous process can be described in five steps. The first step in the process is to prepare the feed, which can be in the form of a solution, dispersion, or emulsion, depending on the polarity of the active ingredient and carrier. Regardless of its constitution, the feed should be in a pumpable form, either as a liquid or paste, since it will be pumped using a peristaltic pump to the atomizer in the second step of the process. The third step involves producing a spray in the drying chamber by converting the feed into droplets using a nozzle, which may be of three types: rotary, pressurized, or two-fluid nozzle. The production of the spray is

Samantha Cristina de Pinho and Carmen S.F. Trindade (eds.), *Bioactives Encapsulation: Food Applications*, Methods and Protocols in Food Science, https://doi.org/10.1007/978-1-0716-4538-3_4,

crucial to ensure a fast and efficient heat exchange in the drying chamber, leading to the evaporation of the water due to contact with the flow of hot air. Inside the drying chamber, droplets are instantly transformed into particles and separated from the air using a cyclone device. The particles are then deposited in a glass collector at the bottom of the cyclone.

Compared to other drying techniques, such as freeze-drying, spray-drying is faster and cheaper due to its less demand for energy and instant dehydration. Powders produced by spray-drying are more fluid, have better wettability, and are less prone to agglomeration compared to those produced by freeze-drying. Indeed, this technique is highly desirable for both laboratory and industrial applications and, despite being a well-established technology, continuous innovations have been made in the field, particularly in terms of carrier materials and equipment.

The process settings in spray-drying can be categorized as either open-loop or closed-loop. The first method uses non-recycled air as a drying gas, while the second method uses a recycled inert gas such as nitrogen throughout the drying process [1].

Spray-drying is a widely used technology for encapsulation in the food industry due to its low cost and the availability of equipment. This technology was initially used in the 1930s to encapsulate aromas using Arabic gum as a carrier [2].

The size distribution of particles obtained through spray-drying is relatively narrow, ranging from submicron to micron scale, depending on the atomization device configuration [1]. The particles are matrix-type and they are usually hollow with the active ingredient spread all over the shell of the particle, which is mainly composed of the carrier (Fig. 1). The solubility of the particle depends on the carrier solubility, but they are generally water soluble. Other types of particles obtained by spray-drying are presented in Fig. 2.

When selecting carrier polymers, it is important to consider a variety of physicochemical properties and characteristics. Some key factors include cost, the ability to form solutions with low viscosity even in high concentrations of solids, solubility, molecular weight, glass/melting transition, crystallinity, diffusibility, film-forming and emulsifying properties, and sensory properties [3].

In the microencapsulation of food ingredients, carrier polymers such as Arabic gum, maltodextrins, pectin, and modified starches are commonly employed. Among these, Arabic gum is considered the most effective natural polymer for encapsulation through spray-drying. It enhances the emulsification properties and film formation while providing excellent water solubility. Maltodextrins are also highly soluble in water but may require an additional emulsifier when encapsulating hydrophobic cores. The use of low molecular weight carbohydrates in microencapsulation is not recommended,

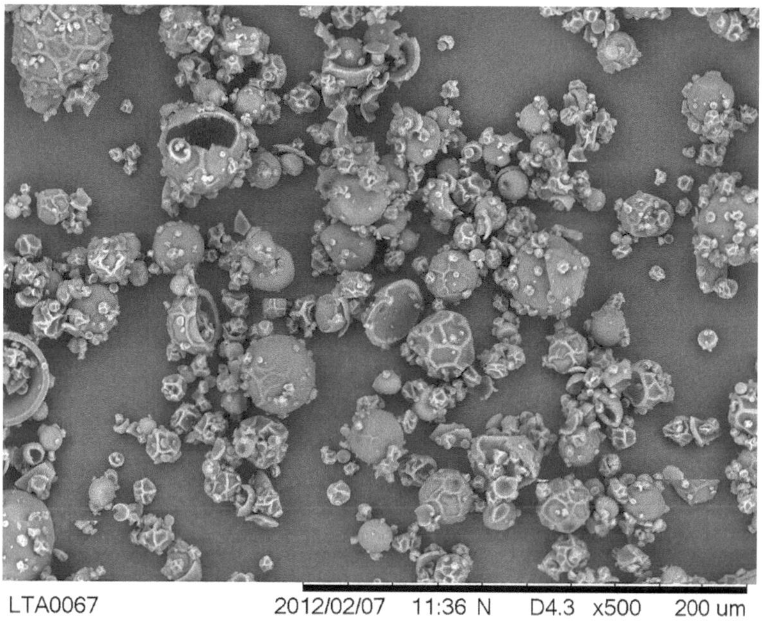

Fig. 1 Spray-dried particles obtained from a solution of grape pomace extract using maltodextrin as carrier. The different sizes of the particles, the spherical shape, and some hollow particles may be observed

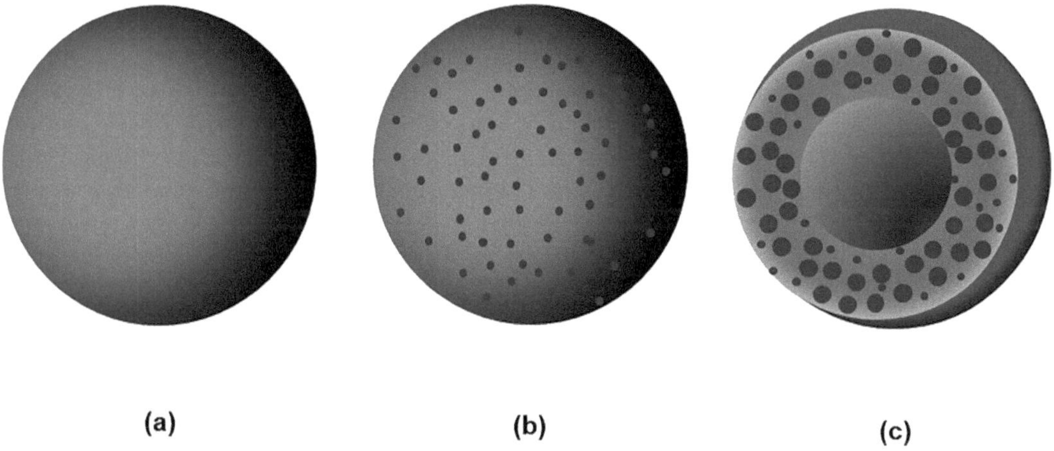

Fig. 2 Types of microparticles/microspheres obtained by spray-drying. (**a**) The feed mixture is a solution. The active ingredient is distributed homogenously throughout the particle structure. (**b**) The feed mixture is a dispersion. The active ingredient is dispersed throughout the particle structure. (**c**) The feed mixture is an emulsion. The droplets of the active ingredient are distributed on the wall of the hollow particle

since they are usually associated with some powder issues, such as caking and re-crystallization of amorphous carbohydrates upon storage. In addition, according to Gharsallaoui et al. [3], the spray-drying technique is well known to be unsuitable for producing powders containing high content of sugar and acid-rich foods due to their stickiness.

Proteins and Arabic gum are classified as amphiphilic molecules, but many proteins are not highly soluble. Some proteinaceous materials, such as whey protein, sodium caseinate, gelatin, and soy protein isolate, are commonly used as carriers for encapsulation by spray-drying. These compounds are known to possess good microencapsulating properties [3].

This technology offers several advantages over conventional techniques for particle production. It features a simple, continuous processing operation, high productivity, and easy scalability. Additionally, it enables the production of powders that are easily manipulated and incorporated into different food matrices, contributing to lower transport and storage costs due to the smaller volume. However, some applications may be hindered by the fact that the particles produced are highly water soluble. In comparison to freeze-dried powders, spray-dried powders have better flow properties. Furthermore, the cost of the spray-drying process is significantly lower, ranging from 30 to 50 times less than freeze-drying [4].

Indeed, spray-drying is a microencapsulation technique that offers several advantages for food compounds due to the protection against degradation, higher stability, better handling of liquid labile active compounds, and higher solubility and dispersibility. In addition, spray-dried microparticles may control the release of active ingredients and mask unpleasant aromas and flavors [5].

Spray-drying is a widely used technique in the industry to dry and process products into powders or granules. However, according to Sosnik & Seremeta [1], the yield of laboratory spray dryers is not optimal when compared to the industrial ones (around 20–70%). This is due to the loss of product that gets stuck in the walls of the drying chamber and the inability of small cyclones to separate fine particles that are less than 2 μm in size. As a result, the yield for spray dryers at both laboratory and plant scales is generally low.

Thus, this protocol is designed to provide an introduction to the spray-drying encapsulation method. It aims to guide the reader through the main steps required for the production of particles using this technology, which can be used to deliver compounds with a wide range of properties.

2 Materials

2.1 Spray-Drying Using Hydrophilic Active Ingredients

2.1.1 Types of Active Ingredients

Materials that are highly soluble or that can be easily dispersed in water are used. Using a dispersion or a solution is possible to entrap or immobilize probiotics [6], vegetable extracts [7], water-soluble natural pigments [8], protein hydrolysates [9, 10], and water-soluble vitamins [11].

2.1.2 Carrier Agents

Some of the most commonly used carriers are maltodextrins with varying degrees of equivalent dextrose [8], Arabic gum [12], and some modified starches [13], etc.

2.1.3 Preparation of the Feed Mixture

(i) Dilute or disperse the active ingredient in distilled water. When it is an aqueous extract, use it directly or dilute it with distilled water if necessary. If it is an alcoholic extract, concentrate in a rotary evaporator at a maximum of 60 °C to remove the alcohol.

(ii) Dilute or disperse the carrier in the mixture containing the active ingredient at the desired concentration under constant stirring. Room temperature or gentle heating is normally used until a homogeneous mixture is obtained. If the feed mixture is a dispersion, keep it under constant agitation during the drying process.

2.2 Spray-Drying Using Hydrophobic Active Ingredients

2.2.1 Types of Active Ingredients

In the food industry, spray-drying technology has been employed for microencapsulation of a huge variety of materials. If an emulsion is used as a feed it is possible to encapsulate oils [14], carotenoids [15], fat-soluble vitamins [16], oleoresins [17], and aromas [18].

2.2.2 Carrier Agents

In this case, the wall material must have the ability to form emulsions with the active material. Some of the materials used are Arabic gum [19] and modified starches [20]. For some wall materials, it may be necessary to add an emulsifying agent to the feed formulation [18].

2.2.3 Preparation of the Feed Formulation

(i) Dilute or disperse the wall material in distilled water at the desired concentration under constant stirring until a homogeneous mixture is obtained.

(ii) Add the active material into the carrier solution. Promote high-speed agitation (minimum 8000 rpm) using a homogenizer until an emulsion is obtained.

3 Methods

3.1 Double-Fluid Atomizing Nozzle Assembly (Fig. 3a) (See also Note 1)

(i) Assemble the atomizing nozzle by fitting the upper connections and the lower part with the desired nozzle opening to the atomizer body.

3.2 Assembling the Spray Dryer (Fig. 3b) (see also Note 1)

(i) Fix the upper part of the drying chamber (1). In some equipment, this step is not necessary.

(ii) Connect the drying chamber (2) to the cyclone (3).

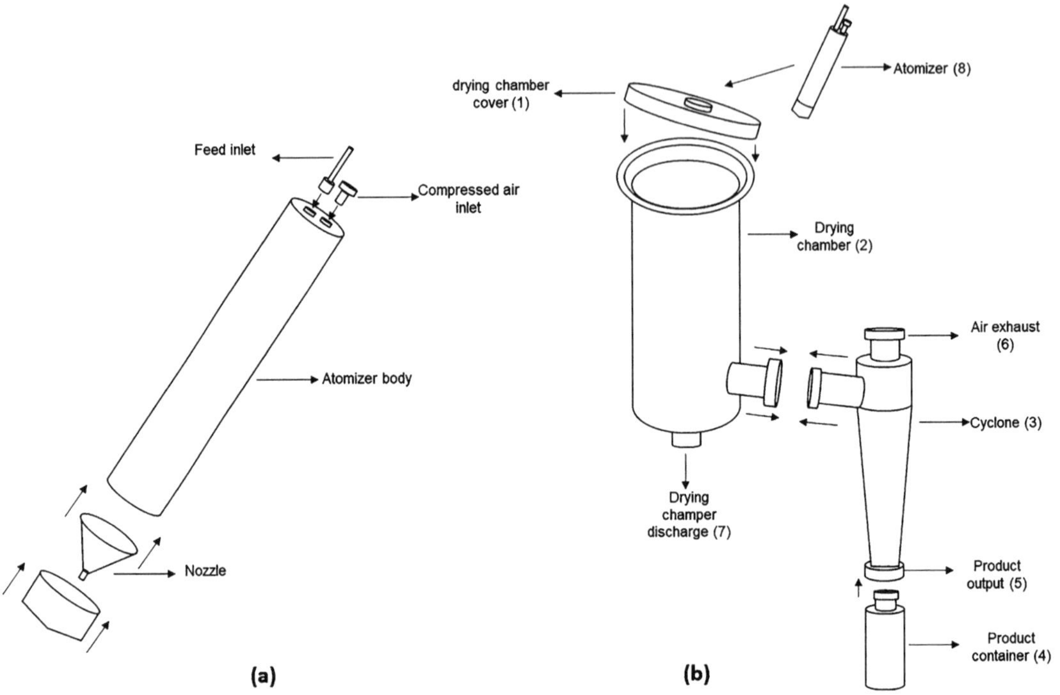

Fig. 3 Equipment assembly diagram. (**a**) Assembling of the double-fluid atomizing nozzle and (**b**) assembling of the spray dryer

(iii) Fix the powder product container (4) in the product output (5) at the bottom of the cyclone. Weigh and record the mass of the empty container to calculate the product yield.

(iv) Connect the upper output of the cyclone (6) to a dust filter or discharge pipe.

(v) Close the drying chamber discharge (7).

(vi) Fix the atomizing nozzle (8) on the top of the drying chamber (1).

(vii) Connect the compressed air circuit to one of the atomizing nozzle inlets (Fig. 3a).

(viii) Connect the feed circuit to the other atomizing nozzle inlet (Fig. 3a). Couple the feed circuit to the equipment's peristaltic feed pump or external feed pump.

(ix) Connect the feed circuit to the feed tank mixture.

3.3 Drying Process (Fig. 4) (see also Note 1)

(i) Turn on the main switch on the equipment panel.

(ii) Turn on the drying air fan on the equipment panel. Adjust the air flow rate to the desired value (*see* **Notes 2** and **3**).

(iii) Turn on the air heating on the equipment panel and set the drying air inlet temperature to the desired value (*see* **Notes 2** and **3**). Depending on the equipment, the sensors are

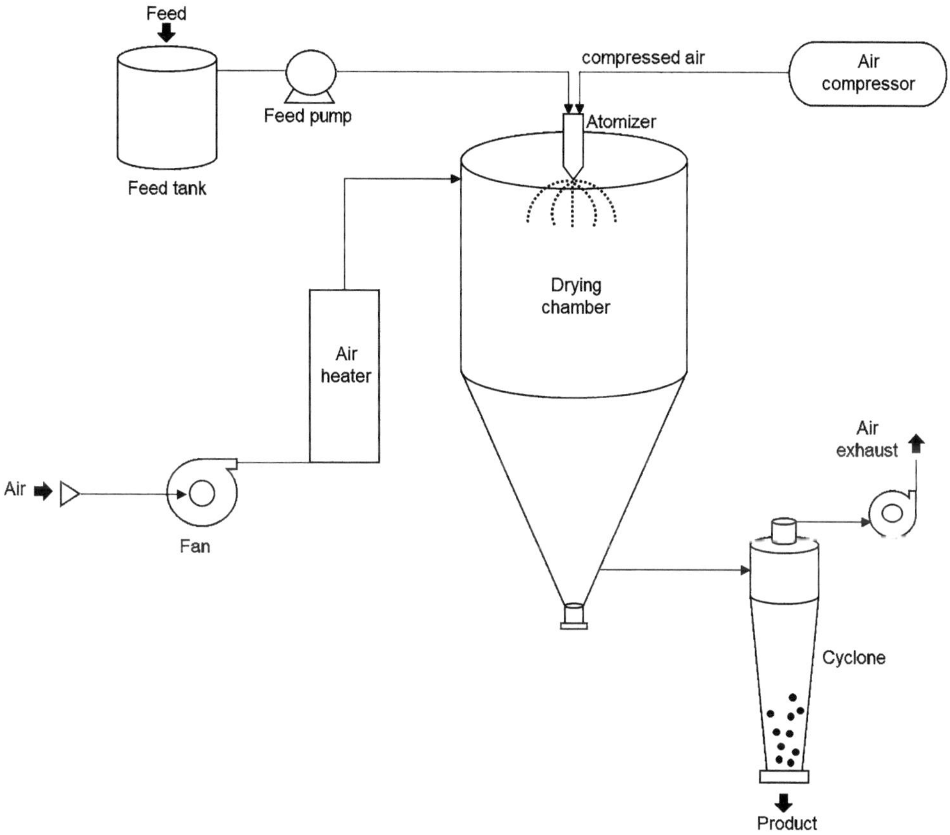

Fig. 4 Drying process diagram using the spray dryer with a double-fluid atomizing nozzle

attached to measure the inlet and output air temperatures. During the process, record the output temperature shown on the equipment.

(iv) After the drying air inlet temperature is reached, turn on the air compressor if the equipment has a double-fluid atomizer nozzle. Adjust the air pressure to the desired value.

(v) Keep the feed mixture under constant stirring if emulsions and dispersions are used. If the feed mixture is a solution, stirring is not mandatory (*see* **Note 4**).

(vi) Adjust the feed flow rate of the peristaltic pump to the desired value (*see* **Note 3**). Turn on the peristaltic pump.

(vii) Before starting to dry the feed mixture, pump distilled water for a few seconds at the beginning of the process, to remove air from the feed circuit (*see* **Note 5**).

(viii) Dry the feed mixture until all liquid is pumped from the feed tank. Finally, pump distilled water for a few seconds to remove the feed mixture that remains stuck inside the feed circuit (*see* **Note 6**).

(ix) Carefully remove the product container (*see* **Notes 6–9**). Weigh the container to determine the product yield.

(x) Turn off the peristaltic pump by interrupting the feed flow.

(xi) Turn off the air compressor and air heating. Wait for the temperature to decrease to at least 60 °C (*see* **Note 2**).

(xii) Turn off the drying air fan and then turn off the equipment's main switch.

(xiii) Store the powder product appropriately for subsequent analyses (*see* **Note 8**). Storage in packaging with a barrier to light, oxygen, and gases may be necessary. Storage at refrigerated, freezing, or room temperatures may be required.

3.4 Disassembling and Cleaning the Equipment

(i) Disassemble the removable parts of the equipment (*see* **Note 6**).

(ii) Clean the internal part of the equipment with water and neutral detergent. Remove all residue of material adhering to the walls of the drying chamber, connections, atomizing nozzle, and cyclone.

(iii) Dry the equipment parts completely before using them again.

4 Notes

1. This protocol is intended for the drying process using laboratory and pilot-scale spray dryer equipment with a double-fluid atomizing nozzle. There may be differences in the assembling and operation of equipment depending on the brand and type of equipment.
2. To avoid damaging the air heater system, never turn on the air heater before turning on the air fan, and never turn off the air fan with the heating turned on and before the temperature drops enough.
3. Process parameters such as inlet air temperature, drying airflow, feed flow rate, and atomizing nozzle opening are variable depending on the type and size of the equipment. These parameters can be adjusted to evaluate the influence of process conditions on product properties.
4. It is recommended to carry out the drying process immediately after preparing the feed mixture. In the case of the feed mixture containing hydrophobic active ingredients, drying must be done as quickly as possible to prevent phase separation between the active ingredient and the wall material.
5. If the feed flow is interrupted during the process, it is recommended to turn off the peristaltic pump to check the problem. With the pump turned off, disconnect the feed circuit from the

inlet of the atomizing nozzle. Wash the internal feed circuit with distilled water to remove possible materials that may be blocking the passage. Reconnect the circuit to the atomizing nozzle, the peristaltic pump, and the feed mixture. Turn the pump on again and restart the drying process.

6. A certain amount of dehydrated product may adhere to the walls of the cyclone. It is recommended to use this product, which can be collected after disassembling the equipment, before washing. If the cyclone is not detachable, tap the outside part of the cyclone with a rubber hammer so that the powder material falls into the collection bottle. Some equipment have a vibrating system or hammers to facilitate the removal of powder adhered to the cyclone.
7. For equipment where there is a powder product output also in the drying chamber, it is not recommended to mix the two products (the drying chamber output product and the cyclone output product), as characteristics such as moisture content and particle size, among others, are different.
8. Particles obtained by spray-drying can be characterized regarding encapsulation efficiency and yield, moisture content, water solubility, hygroscopicity, scanning electron microscopy morphology, size and particle size distribution, stability during storage, and release profile.
9. To avoid accidents, do not touch parts of the equipment without protection while the heating is turned on. Use protective thermal gloves if you need to handle the equipment. Wear a mask to avoid aspiration of dust particles from the product.

References

1. Sosnik A, Seremeta KP (2015) Advantages and challenges of the spray-drying technology for the production of pure drug particles and drug-loaded polymeric carriers. Adv Colloid Interf Sci 223:40–54. https://doi.org/10.1016/j.cis.2015.05.003
2. Shahidi F, Han XQ (1993) Encapsulation of food ingredients. Crit Rev Food Sci Nutr 33(6):501–547. https://doi.org/10.1080/10408399309527645
3. Gharsallaoui A, Roudaut G, Chambin O, Voilley A, Saurel R (2007) Applications of spray-drying in microencapsulation of food ingredients: an overview. Food Res Int 40(9): 1107–1121. https://doi.org/10.1016/j.foodres.2007.07.004
4. Desobry SA, Netto FM, Labuza TP (1997) Comparison of spray-drying, drum-drying and freeze-drying for β-carotene encapsulation and preservation. J Food Sci 62(6): 1158–1162. https://doi.org/10.1111/j.1365-2621.1997.tb12235.x
5. Furuta T, Neoh TL (2021) Microencapsulation of food bioactive components by spray drying: a review. Dry Technol 39(12): 1800–1831. https://doi.org/10.1080/07373937.2020.1862181
6. Oliveira AC, Moretti TS, Boschini C, Baliero JCC, Freitas O, Favaro-Trindade CS (2007) Stability of microencapsulated B. lactis (BI 01) and L. acidophilus (LAC 4) by complex coacervation followed by spray drying. J Microencapsul 24(7):685–693. https://doi.org/10.1080/02652040701532908
7. Daza LD, Fujita A, Fávaro-Trindade CS, Rodrigues-Ract JN, Granato D, Genovese MI (2016) Effect of spray drying conditions on the physical properties of Cagaita (*Eugenia dysenterica* DC.) fruit extracts. Food Bioprod

Process 97:20–29. https://doi.org/10.1016/j.fbp.2015.10.001
8. De Souza VB, Thomazini M, Balieiro JCDC, Fávaro-Trindade CS (2015) Effect of spray drying on the physicochemical properties and color stability of the powdered pigment obtained from vinification byproducts of the Bordo grape (Vitis labrusca). Food Bioprod Process 93:39–50. https://doi.org/10.1016/j.fbp.2013.11.001
9. Favaro-Trindade CS, Santana AS, Monterrey-Quintero ES, Trindade MA, Netto FM (2010) The use of spray drying technology to reduce bitter taste of casein hydrolysate. Food Hydrocoll 24(4):336–340. https://doi.org/10.1016/j.foodhyd.2009.10.012
10. Molina Ortiz SE, Mauri A, Monterrey-Quintero ES, Trindade MA, Santana AS, Favaro-Trindade CS (2009) Production and properties of casein hydrolysate microencapsulated by spray drying with soybean protein isolate. LWT Food Sci Technol 42(5):919–923. https://doi.org/10.1016/j.lwt.2008.12.004
11. Nizori A, Bui LTT, Jie F, Small DM (2020) Spray-drying microencapsulation of ascorbic acid: impact of varying loading content on physicochemical properties of microencapsulated powders. J Sci Food Agric 100(11):4165–4171. https://doi.org/10.1002/jsfa.10455
12. Subtil SF, Rocha-Selmi GA, Thomazini M, Trindade MA, Netto FM, Favaro-Trindade CS (2014) Effect of spray drying on the sensory and physical properties of hydrolysed casein using gum arabic as the carrier. J Food Sci Technol 51(9):2014–2021. https://doi.org/10.1007/s13197-012-0722-z
13. da Silva FC et al (2013) Assessment of production efficiency, physicochemical properties and storage stability of spray-dried propolis, a natural food additive, using gum Arabic and OSA starch-based carrier systems. Food Bioprod Process 91(1):28–36. https://doi.org/10.1016/j.fbp.2012.08.006
14. Xu S et al (2020) Microencapsulation of sea buckthorn (Hippophae rhamnoides L.) pulp oil by spray drying. Food Sci Nutr 8(11):5785–5797. https://doi.org/10.1002/fsn3.1828
15. Rocha GA, Fávaro-Trindade CS, Grosso CRF (2012) Microencapsulation of lycopene by spray drying: characterization, stability and application of microcapsules. Food Bioprod Process 90(1):37–42. https://doi.org/10.1016/j.fbp.2011.01.001
16. Xie Y-L, Zhou H-M, Zhang Z-R (2007) Effect of relative humidity on retention and stability of vitamin a microencapsulated by spray drying. J Food Biochem 31(1):68–80. https://doi.org/10.1111/j.1745-4514.2007.00099.x
17. Ferreira S, Malacrida CR, Telis VRN (2016) Influence of emulsification methods and use of colloidal silicon dioxide on the microencapsulation by spray drying of turmeric oleoresin in gelatin-starch matrices. Can J Chem Eng 94(11):2210–2218. https://doi.org/10.1002/cjce.22596
18. Pilicheva B, Uzunova Y, Katsarov P (2021) Comparative study on microencapsulation of lavender (Lavandula angustifolia Mill.) and peppermint (Mentha piperita L.) essential oils via spray-drying technique. Molecules 26(24): Art. no. 24. https://doi.org/10.3390/molecules26247467
19. de Freitas Santos PD, Rubio FTV, de Carvalho Balieiro JC, Thomazini M, Favaro-Trindade CS (2021) Application of spray drying for production of microparticles containing the carotenoid-rich tucumã oil (*Astrocaryum vulgare* Mart.). LWT 143:111106. https://doi.org/10.1016/j.lwt.2021.111106
20. Partheniadis I, Zarafidou E, Litinas KE, Nikolakakis I (2020) Enteric release essential oil prepared by co-spray drying methacrylate/polysaccharides—influence of starch type. Pharmaceutics 12(6):Art. no. 6. https://doi.org/10.3390/pharmaceutics12060571

Chapter 5

Use of Spray Chilling/Cooling/Congealing/Prilling for the Protection of Nutrients and Bioactive Compounds

Fabricio Luiz Tulini, Volnei Brito de Souza, and Carmen Silvia Favaro-Trindade

Abstract

Spray chilling is a straightforward encapsulation method that involves only two steps to incorporate molecules, microorganisms, or other materials into small particles primarily made from vegetable fats. This technique has gained popularity in the food industry due to the protective structures it produces, which can release their contents under specific conditions, such as during food warming and fat digestion. However, proper carrier selection and temperature control before and during the process are the main challenges for applying spray chilling, as clogging equipment can occur. This protocol aims to guide the reader through the key steps of spray chilling encapsulation, including carrier selection and operational conditions, as well as providing tips for overcoming any challenges that may arise.

Key words Carrier, Encapsulation, Particles, Process

1 Introduction

The process of spray chilling is relatively simple and involves two essential steps. The first step involves the preparation of a liquid feed formulation, which can either be a solution, dispersion, or emulsion, depending on the properties of the active ingredient and carrier. It is important to note that, for food applications, the formulation must contain molten fat as the carrier. The second step involves atomizing the formulation into a cold chamber [1, 2]. To solidify the droplets into spherical solid lipid particles, the atomization chamber must be kept at a temperature lower than the carrier melting point. This is achieved by either implementing a flow of cold air from a laboratory chiller or injecting liquid nitrogen. The heat transfer that takes place in the cooled chamber between the melted carrier and the cold air facilitates the droplets' crystallization, resulting in a powder comprising spherical solid lipid particles. This process is known by several names in the literature, including

Samantha Cristina de Pinho and Carmen S.F. Trindade (eds.), *Bioactives Encapsulation: Food Applications*, Methods and Protocols in Food Science, https://doi.org/10.1007/978-1-0716-4538-3_5,

Fig. 1 Scanning electron micrographs (1500× magnification) of solid lipid microparticles (SLMs) produced with vegetable fat (melting point at 48 °C)

spray cooling, spray congealing, or prilling (the latter being less common). Regardless of the name used, the principle remains the same: producing powders by solidifying droplets into dense solid particles.

Nutrients and bioactive compounds in liquid form can pose challenges in food production due to their high viscosity, low concentration (e.g., vitamins and aromas), instability, high reactivity, and high volatility. To address these issues, microencapsulation techniques have been used for many years to create food ingredients in powder form. The resulting powder, produced through spray chilling, consists of matrix-type particles with the active ingredient dispersed throughout the volume of each particle, including its surface (Fig. 1). These structures are commonly referred to as particles, microparticles, or microspheres, but never microcapsules.

Particles produced by spray chilling are dense, massive, insoluble in water, and with a size range (from nano to micro-sized) governed by the configuration of the device used for atomization [3]. According to Figueiredo et al. [4], solid lipid particles are a fascinating and adaptable means of vehiculating active compounds. Their kinetic stability and rigid structure provide clear advantages over vesicular lipid colloidal systems, such as liposomes and niosomes. To ensure optimal performance and health benefits, it is recommended to use a heterogeneous fat carrier with high solid content and a melting point above room temperature. Solid lipid particles have already been studied for many proposals, such as the encapsulation of probiotics [5], probiotics and prebiotics [6], vitamin C [7, 8], vitamin D3 [9], vitamin B12 [10], oleoresins, lycopene [11], protein hydrolysates [12], antioxidants (as gallic acid, quercetin, curcumin, cinnamon extract, genipap extract, and polyphenols), and minerals (iron and zinc).

Additional research is necessary to assess the impact of using spray-chilled particles in various food and beverage compositions, as it remains a complex field. Also, there is a need for studies on the bioaccessibility and epithelial transport of molecules encapsulated by spray chilling. This technique is primarily applied in the commercial sector for encapsulating bakery ingredients to facilitate their release during bread-making, as well as for aromas to achieve controlled release during chewing or warming and water-insoluble vitamins for release and absorption in the intestine.

Thus, this chapter aims to introduce the spray chilling encapsulation method and guide the reader through the production of solid lipid microparticles as vehicles for different compounds.

2 Materials

Equipment (see **Note 1**):

(i) Laboratory water bath. The equipment must reach at least 10 °C above the melting point of the carrier.

(ii) Disperser with stainless steel dispersing element. The equipment must work above 8000 rpm.

(iii) Spray chiller equipment (*see* **Note 2**).

(iv) A trace heating accessory is important to keep the product above the carrier melting temperature before atomization (*see* **Notes 3** and **4**).

Carriers: Carriers must have a melting point according to the final application of the product. For food applications, carriers must be food grade or present a GRAS status (*see* **Notes 5–9**).

Ingredients to be loaded: Several ingredients may be loaded into the microparticles, such as powders, liquid materials, oleoresins, microorganisms, vitamins, lycopene, protein hydrolysates, antioxidants (gallic acid, quercetin, curcumin, cinnamon extract, genipap extract, and polyphenols), minerals (iron and zinc), and others [10–13].

3 Methods

(i) Heat the carrier 10 °C above the melting point to melt all the material.

(ii) Add the ingredient into the molten carrier and mix at 8000 rpm for 1 min (*see* **Note 10**).

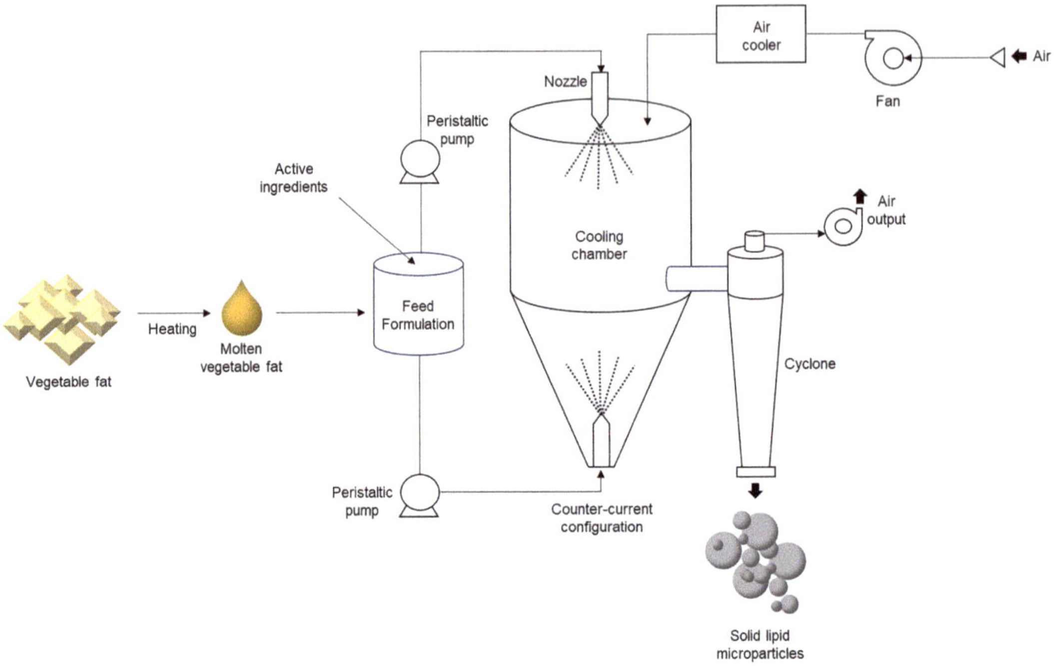

Fig. 2 Example of the process of production of solid lipid microparticles with vegetable fat and using a spray chiller. The counter-current configuration may be set in some equipment

(iii) Set the spray chiller to reach less than 10 °C in the atomization chamber (*see* **Notes 11** and **12**).

(iv) Immediately start the atomization process of the mixture at a flow rate of ca. 60 mL/min. It is important to keep the mixture in constant heating and stirring.

(v) Collect the microparticles produced in the chamber of the spray chiller (*see* **Notes 13** and **14**).

(vi) The microparticles may be analyzed concerning the morphology, chemical structure, and release profile, and further applied in several products (*see* **Notes 15–18**).

(vii) Figure 2 presents an example of this process.

4 Notes

1. This protocol was based on a spray chiller equipped with a laboratory chiller (to chill the inlet air) and a trace heating accessory that keeps the formulation above a certain minimum temperature before atomization. This accessory also heats the pump body and spray nozzle to avoid product solidification before atomization. However, other possibilities for equipment adaptations are presented throughout the protocol.

2. Alternatively, a spray dryer may be used if a laboratory chiller is connected to the air inlet of the equipment [14].
3. As an alternative, a magnetic stirrer capable of reaching at least 10 °C above the melting point of the carrier can be used.
4. If the equipment does not have a system to heat the nozzle body, it can be preheated in a laboratory oven for 1 h.
5. For food applications, where the carrier must be generally recognized as safe (GRAS) or food grade, the most common carriers used for the production of particles by spray chilling are vegetable fats, with high content of solids at room temperature, and with melting points varying from 36 to 60 °C [5, 8, 9, 11, 13, 15].
6. In addition to vegetable fats, other materials can be used as carriers, such as pure fatty acids [16, 17], waxes, such as carnauba wax, candelilla wax, beeswax, solid paraffin, rice bran wax [18], as well as a fully hydrogenated milk fat [19].
7. According to Ref. [20], the formation of a less ordered solid lipid matrix, with different triacylglycerols and fatty acids (with different melting points), favors the accommodation and permanence of the active compound in the lipid matrix.
8. Low-melting lipids such as cocoa butter may cause some problems during and after the spray chilling process. If the chamber is not adequately cooled, it can result in the absence of microparticle formation or high adhesion within the chamber, as well as the formation of particles with high powder adhesion, agglomeration, and low fluidity [5, 12, 21, 22].
9. Lipids with high melting points (above 70 °C) can clog the atomizing nozzle [20]. Several issues arise from utilizing fats with a high melting point. These include the deterioration of the active ingredient in the feed before atomization, the heightened risk of expulsion of the active ingredient from the particles due to fat recrystallization during storage, and the fact that high levels of saturated fats are detrimental to health. Therefore, it is not suggested to employ such fats in food applications.
10. To ensure proper atomization, in the absence of a heating system for the formulation, it is recommended to place the mixture in a magnetic stirrer with a magnetic bar until the atomization process begins. Additionally, it is recommended to set the heating to 10 °C above the carrier melting point.
11. If the equipment lacks a system to heat the nozzle body, insert and fix the pre-heated atomization nozzle into the pre-cooled spray chiller.

12. Some equipment can set the atomization nozzle in a countercurrent configuration, allowing the product to stay in the atomization chamber for a longer time to increase the yield.
13. The process yield can be calculated by dividing the mass of microparticles by the mass of the formulation before atomization.
14. The encapsulation efficiency can be determined by dividing the mass of the active ingredient in the microparticles by the mass of the active ingredient in the formulation.
15. Release profile. The active ingredients within spray-chilled particles added to food can be released through various processes, such as during bread baking or chewing. However, the primary release occurs during fat digestion in the small intestine. In the case of hydrophilic active ingredients that were not emulsified, there are other possibilities of release, such as through diffusion, erosion, and leaching of the matrix due to its hydrophobic nature.
16. Compared to other particle production methods, this technology has several advantages. It is a continuous process that can be easily scaled up, making it faster and highly productive. Additionally, it does not require high temperatures, resulting in greater encapsulation efficiency. This technology also allows for obtaining powders, which makes it easier to manipulate and incorporate into various food matrices. Furthermore, it contributes to lower transport and storage costs due to its smaller volume. Post-processing techniques such as agglomeration and coating are not necessary. Finally, it is a clean technology that does not require the use of solvents and does not produce any residues [1].
17. Moreover, this technology enables the co-encapsulation of multiple ingredients within a single structure. A previous study reported the benefits of co-encapsulation, where the incorporation of proanthocyanidins from cinnamon extract along with α-tocopherol in spray-chilled particles improved the scavenging ability of α-tocopherol [13]. Additionally, the co-encapsulation of probiotics and prebiotics resulted in the creation of symbiotic structures [6].
18. On the other hand, there are some drawbacks associated with particles produced by the spray chilling process. For instance, if the carrier fat recrystallizes, the particles may expel their active ingredients during storage. Moreover, as they are highly hydrophobic, they cannot be applied to all types of food. Additionally, the concentration of active ingredients in the feed must be kept relatively low, which is contingent on numerous factors, including the final viscosity of the feed and whether the core is solid or liquid. Nevertheless, it is challenging to exceed a concentration of more than 15%.

Acknowledgments

Carmen S. Favaro-Trindade thanks Conselho Nacional de Desenvolvimento Científico e Tecnológico (CNPq) for the productivity grant (#310686/2022-9).

References

1. Favaro-Trindade CS, Okuro PK, Matos FE Jr (2015) Encapsulation via spray chilling/cooling/congealing. In: Mishra M (ed) Handbook of encapsulation and controlled release. CRC Press, Boca Raton, pp 71–87
2. Okuro PK, Matos Junior FE, Fávaro-Trindade CS (2013) Technological challenges for spray chilling encapsulation of functional food ingredients. Food Technol Biotechnol 51(2): 171–182
3. Favaro-Trindade CS et al (2021) Encapsulation of active pharmaceutical ingredients in lipid micro/nanoparticles for oral administration by spray-cooling. Pharmaceutics 13(8): 1186. https://doi.org/10.3390/pharmaceutics13081186
4. Figueiredo J d A et al (2023) Microstructured lipid microparticles containing anthocyanins: production, characterization, storage, and resistance to the gastrointestinal tract. Food Res Int 166:112611. https://doi.org/10.1016/j.foodres.2023.112611
5. Pedroso DL, Dogenski M, Thomazini M, Heinemann RJB, Favaro-Trindade CS (2013) Microencapsulation of Bifidobacterium animalis subsp. lactis and Lactobacillus acidophilus in cocoa butter using spray chilling technology. Braz J Microbiol 44(3):777–783. https://doi.org/10.1590/S1517-83822013000300017
6. Okuro PK, Thomazini M, Balieiro JCC, Liberal RDCO, Fávaro-Trindade CS (2013) Co-encapsulation of Lactobacillus acidophilus with inulin or polydextrose in solid lipid microparticles provides protection and improves stability. Food Res Int 53(1):96–103. https://doi.org/10.1016/j.foodres.2013.03.042
7. Matos-Jr FE, Comunian TA, Thomazini M, Favaro-Trindade CS (2017) Effect of feed preparation on the properties and stability of ascorbic acid microparticles produced by spray chilling. LWT 75:251–260. https://doi.org/10.1016/j.lwt.2016.09.006
8. Matos-Jr FE, Di Sabatino M, Passerini N, Favaro-Trindade CS, Albertini B (2015) Development and characterization of solid lipid microparticles loaded with ascorbic acid and produced by spray congealing. Food Res Int 67:52–59. https://doi.org/10.1016/j.foodres.2014.11.002
9. Paucar OC, Tulini FL, Thomazini M, Balieiro JCC, Pallone EMJA, Favaro-Trindade CS (2016) Production by spray chilling and characterization of solid lipid microparticles loaded with vitamin D. Food Bioprod Process 100: 344–350. https://doi.org/10.1016/j.fbp.2016.08.006
10. Mazzocato MC, Thomazini M, Favaro-Trindade CS (2019) Improving stability of vitamin B12 (Cyanocobalamin) using microencapsulation by spray chilling technique. Food Res Int 126:108663. https://doi.org/10.1016/j.foodres.2019.108663
11. Pelissari JR et al (2016) Production of solid lipid microparticles loaded with lycopene by spray chilling: structural characteristics of particles and lycopene stability. Food Bioprod Process 98:86. https://doi.org/10.1016/j.fbp.2015.12.006
12. Salvim MO, Thomazini M, Pelaquim FP, Urbano A, Moraes ICF, Favaro-Trindade CS (2015) Production and structural characterization of solid lipid microparticles loaded with soybean protein hydrolysate. Food Res Int 76: 689–696. https://doi.org/10.1016/j.foodres.2015.08.003
13. Tulini FL et al (2017) Evaluation of the release profile, stability and antioxidant activity of a proanthocyanidin-rich cinnamon (*Cinnamomum zeylanicum*) extract co-encapsulated with α-tocopherol by spray chilling. Food Res Int 95:117. https://doi.org/10.1016/j.foodres.2017.03.010
14. Consoli L, Grimaldi R, Sartori T, Menegalli FC, Hubinger MD (2016) Gallic acid microparticles produced by spray chilling technique: Production and characterization. LWT Food Sci Technol 65:79–87. https://doi.org/10.1016/j.lwt.2015.07.052
15. Marinho JFU, Mazzocato MC, Tulini FL, Silva MP, de Martinis ECP, Fávaro-Trindade CS (2019) Evaluation of probiotic and synbiotic jussara sorbets. Nutr Food Sci 50(2):373. https://doi.org/10.1108/NFS-03-2019-0102

16. Chambi HNM, Alvim ID, Barrera-Arellano D, Grosso CRF (2008) Solid lipid microparticles containing water-soluble compounds of different molecular mass: production, characterisation and release profiles. Food Res Int 41(3): 229–236. https://doi.org/10.1016/j.foodres.2007.11.012
17. Sartori T, Consoli L, Hubinger MD, Menegalli FC (2015) Ascorbic acid microencapsulation by spray chilling: production and characterization. LWT Food Sci Technol 63(1):353–360. https://doi.org/10.1016/j.lwt.2015.03.112
18. Yin Y, Cadwallader KR (2019) Spray-chilling encapsulation of 2-acetyl-1-pyrroline zinc chloride using hydrophobic materials: storage stability and flavor application in food. Food Chem 278:738–743. https://doi.org/10.1016/j.foodchem.2018.11.122
19. Queirós M d S, Viriato RLS, Ribeiro APB, Gigante ML (2020) Dairy-based solid lipid microparticles: a novel approach. Food Res Int 131:109009. https://doi.org/10.1016/j.foodres.2020.109009
20. Müller RH, Radtke M, Wissing SA (2002) Nanostructured lipid matrices for improved microencapsulation of drugs. Int J Pharm 242(1–2):121–128. https://doi.org/10.1016/S0378-5173(02)00180-1
21. Procopio FR et al (2018) Solid lipid microparticles loaded with cinnamon oleoresin: Characterization, stability and antimicrobial activity. Food Res Int 113:351–361. https://doi.org/10.1016/j.foodres.2018.07.026
22. Oriani VB, Alvim ID, Consoli L, Molina G, Pastore GM, Hubinger MD (2016) Solid lipid microparticles produced by spray chilling technique to deliver ginger oleoresin: Structure and compound retention. Food Res Int 80:41–49. https://doi.org/10.1016/j.foodres.2015.12.015

Chapter 6

Microencapsulation of Bioactive-Rich Materials by the Combination of Spray Drying and Spray Chilling

Samuel Henrique Gomes de Sá, Priscilla Dayane de Freitas Santos, and Carmen Silvia Favaro-Trindade

Abstract

Microencapsulation technologies have been extensively applied in the food industry during the last decades, due to their many advantages, such as protecting the core from degradation and converting liquid feeds into powdered materials. Spray drying and spray chilling are widespread processes to produce particles in the form of powders, in the micron scale, and can be used to encapsulate hydrophobic and hydrophilic compounds. However, both techniques present disadvantages that can limit the protection of the core and their application in foods, e.g., exposure to high temperatures during spray drying and polymorphic changes of the wall material in spray-chilled particles. A promising alternative to overcome these drawbacks is the combination of spray drying and spray chilling to produce double-coated bioactive compounds, thus adding up the benefits of both processes, as well as mitigating their limitations. Recent studies that applied this approach to encapsulate different core materials, i.e., fish oil, probiotics, tucumã oil, red propolis extract, phenolic compounds, and carotenoid-rich extracts from pumpkin and guarana by-products, demonstrated that the combination of spray drying and spray chilling provided additional protection to the encapsulated compounds by reducing the loss of bioactive compounds after heating and/or storage, besides allowing to mask unpleasant flavors and improving the controlled release of the core during simulated gastrointestinal digestion. Based on these successful results, this protocol aims to give detailed information about the basic steps to produce microparticles by the combination of spray drying and spray chilling, in addition to providing key notes and tips for adaptation of the method to different core materials and for troubleshooting.

Key words Double coating, Microparticles, Carotenoids, Bioactive compounds, Stability, Controlled release

1 Introduction

Microencapsulation processes comprise a range of physical, chemical, and physicochemical techniques applied to produce solid or liquid particles of micrometric size. The principle of encapsulation is to imprison a substance inside a layer or matrix of another material, providing a physical barrier between the inner compound

Samantha Cristina de Pinho and Carmen S.F. Trindade (eds.), *Bioactives Encapsulation: Food Applications*, Methods and Protocols in Food Science, https://doi.org/10.1007/978-1-0716-4538-3_6,

and the environment [1, 2]. The encapsulated material can be a solid, liquid, or gas and is referred to as core, active, fill, or internal phase, while the substance used to encapsulate it is named carrier, encapsulant, wall material, shell, or external phase. Both the encapsulated and encapsulating components can be composed of a single ingredient or a mixture [3]. The characteristics and properties of the obtained microparticles or microcapsules, such as structure, morphology, and size, depend on the materials and techniques applied for encapsulation. For example, the structure of microcapsules, also called reservoir structures, varies from a simple core covered by a layer of encapsulant to capsules composed of multiple cores and/or layers. Another type of structure includes microparticles of spherical or irregular shape in which the encapsulated compound is dispersed within a continuous matrix of encapsulant, without forming a proper core [2, 4].

The main purpose of most encapsulation processes is to protect the core from degrading factors and increase its stability. For this reason, microencapsulation of extracts and oils is often employed to prevent losses of sensitive bioactive like carotenoids [5–8], flavonoids [9–12], and phenolic compounds [13–15]. Other advantages of encapsulation include easier handling by conversion of liquids into solids; masking of unpleasant flavors; increased retention of volatiles; controlled release of the core during food storage, process, or digestion; enhancing the dispersibility of hydrophobic compounds in water; and improved bioavailability of drugs and bioactive compounds [1, 3, 16].

Several types of microencapsulation technologies can be used to encapsulate bioactive-rich extracts and oils. Spray drying stands out as the most explored method at the research and industrial levels, due to its simple, fast, continuous, and low-cost operation, together with the high equipment availability and ease of scale-up. This process consists of the atomization of a liquid mixture containing the core and carrier materials inside a chamber with hot air circulation, which promotes instant drying of the droplets and formation of water-soluble powders. Despite its advantages, spray drying results in hollow particles with cracked surface, characteristics that can reduce the protection of the encapsulated compound, and the use of high temperatures, even for a short time, can trigger the degradation of thermolabile bioactive compounds [4, 17, 18].

Spray chilling is a very similar process that addresses the mentioned issues by producing hydrophobic particles with continuous surface at low temperatures. The method is based on incorporating the core into a molten lipid followed by atomization of the solution or emulsion inside a cold chamber (usually at temperatures lower than 15 °C), which results in solidification of the lipidic carrier and formation of solid lipid microparticles [19]. Some limitations of spray chilling include potential polymorphic changes or oxidation of the lipid carrier and reactions that can decrease the protection and stability of the core [20].

Considering that any encapsulation method presents specific disadvantages, it was suggested, decades ago, that the combination of different techniques could be a way of overcoming their limitations and bringing additional benefits to the encapsulated compound [21, 22]. As an example, Rosenberg and Hills [23] described a process in which vitamin A was spray dried and then encapsulated by spray chilling to ensure its protection against prooxidant agents.

Similar approaches were reported again in the literature only a few years ago, for encapsulation of fish oil (rich in omega-3) and probiotics (*Saccharomyces boulardii*, *Lactobacillus acidophilus*, and *Bifidobacterium bifidum*), by the combination of spray drying and spray chilling [24–28]. More recently, other authors studied the application of spray drying followed by spray chilling to produce microparticles loaded with red propolis extract and carotenoid-rich extracts from pumpkin and guarana by-products [29–31]. Fadini et al. [28] observed that the double coating masked the unpleasant flavor of fish oil, while particles obtained only by spray drying had a noticeable off-flavor during sensory acceptance analysis. According to Sá et al. [31] and Lima et al. [29], the combination of methods improved the release profile of encapsulated bioactive compounds compared to materials produced by the individual processes, resulting in gradual release of formononetin and carotenoids, respectively, during simulated digestion of the microparticles. Lima et al. [29] and Pinho et al. [30] also evidenced that combining spray drying and spray chilling provided higher oxidative stability to pumpkin and guarana extracts than spray chilling alone, considering that oxidation induction times of the samples exposed to heating at 120 °C increased by approximately 198–253-fold after the double coating.

In a recently published paper, it was shown that tucumã oil microparticles obtained by spray drying followed by spray chilling had lower carotenoid degradation (about 5% after 90 days of storage) compared to particles produced by spray drying alone (approximately 21% of carotenoid loss at the end of the same period) [32]. In another work, it was demonstrated that the encapsulation of a carotenoid-rich extract, obtained from pumpkin peels, by spray drying followed by spray chilling protected the carotenoids from degradation upon extrusion cooking of cornmeal. The authors observed that the combined methods were more effective in preserving both α- and β-carotene after extrusion (≈ 43–50% of retention) in comparison to spray drying (about 25–35% of retention) or spray chilling alone (approximately 25% of retention for both carotenoids) [33]. Another recent study showed that microparticles produced by the combination of spray drying and spray chilling presented controlled release of phenolic compounds, such as chlorogenic acid and myricetin, during gastrointestinal digestion [34].

The results of these works reveal that the encapsulation of bioactive compounds and oxidation sensitive materials by spray drying followed by spray chilling is a promising strategy to produce particles with improved sensory, release, and stability properties, and that further studies are needed to investigate the impact of this approach on other bioactive-rich materials. In addition, the application of these microparticles to food and packaging formulations should also be studied, to evaluate their effectiveness in protecting the encapsulated compound from harsh processing conditions, e.g., during baking and extrusion, and their effects on the properties of the enriched materials.

Therefore, the aim of this protocol is to provide guidance on the basic procedures for microencapsulation of diverse core materials by the combination of spray drying and spray chilling. The protocol below is based on recent studies to provide examples of formulations that can be used to produce coated particles by combining spray drying and spray chilling techniques, as well as examples of equipment and process conditions at the laboratory level.

2 Materials

- Core material in solution or solid state (e.g., bioactive-rich extracts and oils, probiotics in powdered form, and dried particles produced by spray drying, among others);
- Carrier material for spray drying, e.g., gum Arabic, maltodextrin, or modified starches;
- Carrier material for spray chilling: vegetable fats with a melting point of 32–55 °C, e.g., hydrogenated and/or interesterified vegetable fats (*see* **Note 1**).

3 Methodology

The double-layer encapsulation of bioactive-rich materials using atomization methods requires an initial encapsulation step by spray drying, and then the resulting particles, in powdered form, are subjected to the spray chilling process. A schematic representation of the combination of spray drying and spray chilling is presented in Fig. 1.

(i) Gum Arabic, maltodextrin, and modified starches are the most commonly used carrier agents for the spray drying of hydrophobic and hydrophilic core materials. Gum Arabic and some modified starches present emulsifying capacity, so they are

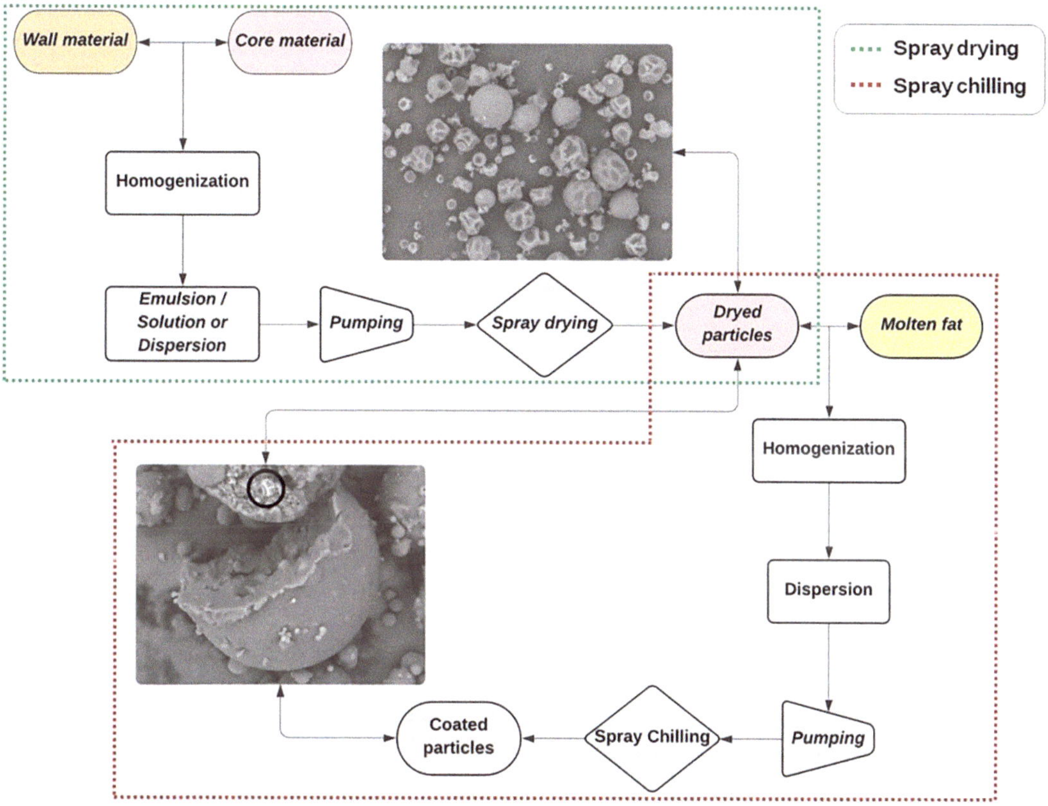

Fig. 1 The production process of coated particles by combining spray drying and spray chilling methods

3.1 Phase 1: Spray Drying

3.1.1 Carrier Material Election and Preparation

more appropriate carriers for encapsulation of hydrophobic compounds, such as vegetable oils, since they increase the stability of the feed fluid (in this case, an emulsion), and the coating of the core, which improves the encapsulation process effectiveness. As well as Arabic gum, maltodextrin produces feed solutions with low viscosity, even at high concentrations, but does not possess emulsifying ability, so it is suitable for encapsulation of hydrophilic cores or an emulsifier should be added to hydrophobic compound encapsulation. Modified starches present varying properties and applicabilities, depending on the applied modification [35]. Proteins also can be used for spray drying of compounds with different characteristics, owing to their amphiphilic nature. However, on the contrary of maltodextrin and Arabic gum, in general, proteins form solutions of high viscosity in low concentration, what is not so appropriate for pumping and nebulization. In addition, feeds with low solid concentrations result in high cost of process and low process yield.

(ii) Industrial and laboratory-scale equipment have different limitations depending on their models and operating capabilities. Simpler equipment is generally capable of atomizing feed solutions with up to 30% dry matter, while more concentrated solutions can cause clogging of the atomizing nozzle *(see* **Note 2**).

(iii) To prepare a solution with 30% of the encapsulating agent, weigh 30 g of the chosen material in powder form in a 250 mL beaker, add 70 g of distilled water, and homogenize with Ultra-turrax ® IKA T25 (Labortechnik, Staufen, Germany) at 10,000 rpm for 3 min or until the solution is completely homogeneous (*see* **Tips 1** and **2**).

3.1.2 Preparation of the Spray Drying Feed Fluid

The feed fluid for spray drying encapsulation is a mixture (solution, emulsion, or a dispersion) of the carrier solution and the core material. The selection of the core:carrier ratio must follow some criteria. Even though a high core proportion in the dried particles is often desired, high core-carrier ratios can result in low encapsulation efficiency and reduced protection of the core. This happens because there should be enough concentration of carrier material in the feed fluid so that the core can be properly embedded within the carrier matrix during the encapsulation process [35]. A core:carrier ratio of 1:4 (20% of core in the dried particles) is frequently reported in published works and generally accepted as an optimal ratio to produce particles using different core materials and carrier agents [17].

(i) Example of feed fluid composition: the 1:4 ratio of bioactive extract (15% dry matter) and wall material solution (30% dry matter) (*see* **Note 3**) results in a feed solution with approximately 27% of total solids (see **Note 4**).

(ii) To produce a feed fluid with a 1:4 ratio: Add 80 g of the 30% (d.m.) encapsulant solution to a 250 mL beaker. Then, add 20 g of the 15% (d.m.) bioactive extract in the solution and homogenize this mixture using an ultraturrax at 10,000 rpm for 3 min or until the solution is homogeneous.

3.1.3 Spray Drying

Recent studies of our group of research used a spray dryer model MSD 1.0 (Labmaq do Brazil Ltda, Ribeirão Preto, Brazil), which is also adapted to work as a spray chiller, to encapsulate different core materials by the combination of spray drying and spray chilling techniques. Table 1 shows the process parameters used by the authors Lima et al. [29], Pinho et al. [30], Sá et al. [31], and Santos et al. [32] (*see also* **Note 5**).

Some tips for the spray drying:

(i) Drain the water buildup in the compressor before starting the process. Atomizing air with high humidity can result in

Table 1
Example of operating conditions for obtaining particles by spray drying

Inlet air temperature	120–180 °C
Outlet air temperature	90–140 °C
Drying air speed	2.50 m/s
Feed flow rate	10 mL/min
Air flow rate	30 L/min
Atomizing nozzle diameter	0.5–1.2 mm

particles with high moisture content and water activity, which might favor the occurrence of chemical and biochemical reactions, or the growth of microorganisms;

(ii) Assemble the spray dryer and set the process parameters, i.e., inlet air temperature, outlet air temperature (*see* **Troubleshooting 1**), drying air speed, feed flow rate, and air flow rate. Examples of common operation parameters are shown in Table 1;

(iii) Atomize the feed fluid right after its preparation, to avoid phase separation;

(iv) If the feed is a dispersion, keep it under magnetic stirring during all the atomization process, to ensure feed homogenization and avoid clogging of the atomizing nozzle (*see* **Troubleshooting 2**);

(v) After all the feed fluid has been atomized, turn off the pump and the heater, and wait until the equipment has cooled down before collecting the dried particles;

(vi) The powders obtained at the end of the spray drying process might be stored at room temperature; however, to avoid some chemical reactions, it is common to be stored at freezing temperatures (−20 °C), protected from ambient moisture and light, and for the shortest period possible.

3.2 Phase 2: Coating of Spray-Dried Particles by Spray Chilling

The core used in this stage consists of spray-dried particles.

3.2.1 Preparation of the Core

(i) To coat the spray-dried particles, it is not necessary to subject them to any additional procedure prior to the second encapsulation process. However, the powder must be stored properly before use (*see* **Note 6**).

3.2.2 Wall Material Preparation

(i) The wall material chosen to coat the spray-dried particles should be a lipid material with high melting point, such as interesterified or a hydrogenated vegetable fat (*see* **Note 7**), besides being food grade and capable of forming solid particles at the selected operating conditions and the available equipment (*see* **Note 8**).

(ii) Using a vegetable fat (with a melting point of 55 °C) as wall material: weigh 100 g of the fat in a 250 mL beaker. Then, place the beaker in a water bath at 70 °C until the material is completely melted (*see* **Note 9**). As the fat becomes translucent or slightly yellow (the color will depend on the composition), which is an indicator of complete melting, it must remain at 65 °C for at least 30 more minutes, to erase the crystalline memory of fat crystals (*see* **Note 10**).

3.2.3 Preparation of the Spray Chilling Feed Fluid

In the proposed double-coating process, the feed fluid is composed of spray-dried particles blended with a molten vegetable fat (wall material), resulting in a dispersion. Once again, the core-to-wall material ratio must be properly chosen to provide as high loading of core as possible but without reducing too much the wall material proportion, which may compromise the efficient coating of the core. In addition, the potential clogging of the atomizing nozzle is more critical in the second encapsulation step, since the core material is a powder, and the viscosity of the feed fluid can increase very quickly if an adequate temperature is not maintained. For atomization in the equipment mentioned above, the maximum proportion of core efficiently incorporated into the molten fat to produce a feed fluid was 20% (w/w) (1:4 ratio) [29–32].

(i) To produce a feed fluid with a core:wall material ratio of 1:4: weigh 80 g of the molten vegetable fat in a 150 mL beaker. Then, add 20 g of the powder obtained by spray drying and homogenize the mixture at 10,000 rpm for 1 min or until the dispersion is homogeneous (*see* **Tip 3**).

(ii) During homogenization, keep the feed fluid in a hot bath to prevent the fat from solidifying (*see* **Note 11**).

Table 2
Example of operating conditions for obtaining particles by the combination of spray drying and spray chilling

Inlet air temperature	13–14 °C
Cooling air speed	1.80 m/s
Feed flow rate	40 mL/min
Atomizing air pressure	1.0 kgf/cm^2
Atomizing nozzle diameter	1.2 mm

3.2.4 Production of Coated Particles

Recent studies combining spray drying and spray chilling techniques managed to produce coated particles using the parameters are described in Table 2 [29–32].

For the coating step by spray chilling:

(i) Determine the process parameters from preliminary trails, considering they depend on the coated compound, the lipid material chosen as encapsulating agent, and the available equipment.

(ii) Keep the pumping hose and atomizing nozzle in an oven at 70 °C (or in the same temperature used for melting of the wall material) for at least 1 h before atomization. However, in some equipment, it is not necessary to pump the feed, which is entering the atomizing nozzle by gravity. Indeed, in this case, it is not necessary to warm up the pumping hose.

(iii) Remove the components from the oven and assemble them in the equipment immediately after homogenizing the feed fluid, and then proceed with pumping the dispersion as soon as possible to prevent clogging of the nozzle (see **Troubleshooting 3**).

(iv) After all the feed fluid has been atomized, collect the coated particles as fast as possible (to avoid partial melting of the particles due to exposure to a potentially high room temperature) and store them at freezing temperatures (−20 °C), protected from light and moisture (*see* **Note 12**).

4 Tips

1. Adding the water over the powdered carrier material used for spray drying prevents foaming during homogenization. Use a spoon to check for lumps after homogenization and, if necessary, continue the process until the lumps are completely dissolved.
2. It is also possible to solubilize the carrier material in water using a magnetic stirrer or orbital shaker. In these cases, the solution must be stirred for 1–2 h and then stored overnight under refrigeration to ensure full hydration of the carrier molecules.
3. It is necessary to heat all the materials that come into contact with the molten fat, such as the homogenizer dispersing tool, the pumping hose, and the atomizing nozzle. These devices must be kept at temperatures close to that of the molten fat, around 70 °C in this example, by the time of contact between them.

5 Notes

1. It is recommended to use a fat with melting point from 32 to 55 °C as wall material during spray chilling, so that the temperature needed for complete melting is not too high (around 65 °C) and to obtain lipid particles that remain solid at room temperature.
2. The maximum concentrations of solids that can be pumped into the spray dryer chamber depend on the type and power of the pump, in addition to the model and diameter of the available atomizing nozzle.
3. If the core material is an extract with low solid content, the solution must be dried or concentrated before adding it to the carrier solution.
4. The solid content of the wall material solution can be adjusted to obtain a feed fluid with total solids above 27%, as long as it does not cause clogging of the atomizing nozzle.
5. Preliminary tests are necessary to determine the process parameters according to the limitations of the equipment, the thermal sensitivity of the core material, and the desired characteristics of the particles.
6. Excess moisture in the internal phase will not be removed during coating by spray chilling. It may result in particles with high moisture content and water activity, which might favor the occurrence of chemical and biochemical reactions, or the growth of microorganisms, thus promoting the degradation of the encapsulated compound.
7. For food applications, the fat used as wall material must have an adequate melting temperature, e.g., melting temperature higher than the room temperature but not so high. Carriers with low melting temperatures require storage of particles under refrigeration, while carriers with high melting temperatures result in more stable particles; however, the consumption of high amounts of saturated fatty acid is not recommended for health. In the same way, for food applications, the use of carriers with trans fatty acid is not recommended.
8. The formation of the coated particles by spray chilling will depend on the melting point of the chosen wall material and the selected process parameters. Preliminary tests will be necessary to determine the most appropriate parameters for each material.
9. Make sure to avoid contact between the fat and the water, to prevent any modification in the fat properties.

10. Melting the wall material 10–15 °C above its melting point helps to ensure that no molecules remain crystallized in the mixture. Furthermore, the additional heating prevents the fat from crystallizing when in contact with the core material and over exposure to the room temperature.
11. Carry out the homogenization of the feed fluid and atomization steps as quickly as possible to avoid loss of thermosensitive compounds and crystallization of the carrier.
12. The storage temperature of the coated particles may vary according to the melting temperature of the wall material.

Trobleshooting

1. Some spray dryers do not allow setting a specific outlet air temperature; in this case, take note of the temperature displayed in the equipment panel.
2. If the atomizing nozzle clogs during the atomization step, one could probably see an amount of feed fluid leaking out at the top of the nozzle, in the junction between the nozzle and the pumping hose. In this case, stop the feed pumping, remove the atomizing nozzle from the spray dryer chamber, and try to unclog it with the aid of compressed air.
3. If the atomizing nozzle clogs during the coating step by spray chilling, once again one could probably see an amount of feed fluid leaking out at the top of the nozzle, in the junction between the nozzle and the pumping hose. In this case, proceed with the steps described in the topic above. In addition, after unclogging the nozzle, return it to the oven at 70 °C for a few minutes before restarting the atomization.

References

1. Gibbs BF, Kermasha S, Alli I, Mulligan CN (1999) Encapsulation in the food industry: a review. Int J Food Sci Nutr 50:213–224. https://doi.org/10.1080/096374899101256
2. Tolve R, Galgano F, Caruso MC, Tchuenbou-Magaia FL, Condelli N, Favati F, Zhang Z (2016) Encapsulation of health-promoting ingredients: applications in foodstuffs. Int J Food Sci Nutr 67:888–918. https://doi.org/10.1080/09637486.2016.1205552
3. Lee, S.J.; Wong, M. Nano- and microencapsulation of phytochemicals.; 2014
4. Gharsallaoui A, Roudaut G, Chambin O, Voilley A, Saurel R (2007) Applications of spray-drying in microencapsulation of food ingredients: an overview. Food Res Int 40:1107–1121. https://doi.org/10.1016/j.foodres.2007.07.004
5. Alehosseini A, Gómez-Mascaraque LG, Ghorani B, López-Rubio A (2019) Stabilization of a saffron extract through its encapsulation within electrospun/electrosprayed zein structures. LWT 113:108280. https://doi.org/10.1016/j.lwt.2019.108280
6. Gomez-Estaca J, Comunian TA, Montero P, Ferro-Furtado R, Favaro-Trindade CS (2016) Encapsulation of an astaxanthin-containing lipid extract from shrimp waste by complex coacervation using a novel gelatin-cashew gum complex. Food Hydrocoll 61:155–162. https://doi.org/10.1016/j.foodhyd.2016.05.005

7. Gomez-Estaca J, Comunian TA, Montero P, Favaro-Trindade CS (2018) Physico-chemical properties, stability, and potential food applications of shrimp lipid extract encapsulated by complex coacervation. Food Bioprocess Technol 11:1596. https://doi.org/10.1007/s11947-018-2116-3
8. Tirado DF, Palazzo I, Scognamiglio M, Calvo L, Della Porta G, Reverchon E (2019) Astaxanthin encapsulation in ethyl cellulose carriers by continuous supercritical emulsions extraction: a study on particle size, encapsulation efficiency, release profile and antioxidant activity. J Supercrit Fluids 150:128–136. https://doi.org/10.1016/j.supflu.2019.04.017
9. Ghatak D, Iyyaswami R (2019) Selective encapsulation of quercetin from dry onion peel crude extract in reassembled casein particles. Food Bioprod Process 115:100–109. https://doi.org/10.1016/j.fbp.2019.03.003
10. Ostroschi LC, Brito de Souza V, Echalar-Barrientos MA, Tulini FL, Comunian TA, Thomazini M, Baliero JCC, Roudaut G, Genovese MI, Favaro-Trindade CS (2018) Production of spray-dried Proanthocyanidin-rich cinnamon (Cinnamomum Zeylanicum) extract as a potential functional ingredient: improvement of stability, sensory aspects and technological properties. Food Hydrocoll 79:343–351. https://doi.org/10.1016/J.FOODHYD.2018.01.007
11. Souza VB, Thomazini M, Echalar Barrientos MA, Nalin CM, Ferro-Furtado R, Genovese MI, Favaro-Trindade CS (2018) Functional properties and encapsulation of a proanthocyanidin-rich cinnamon extract (Cinnamomum Zeylanicum) by complex Coacervation using gelatin and different polysaccharides. Food Hydrocoll 77:297–306. https://doi.org/10.1016/j.foodhyd.2017.09.040
12. Tulini FL, Souza VB, Echalar-Barrientos MA, Thomazini M, Pallone EMJA, Favaro-Trindade CS (2016) Development of solid lipid microparticles loaded with a proanthocyanidin-rich cinnamon extract (Cinnamomum Zeylanicum): potential for increasing antioxidant content in functional foods for diabetic population. Food Res Int 85:10–18. https://doi.org/10.1016/j.foodres.2016.04.006
13. Calomeni AV, de Souza VB, Tulini FL, Thomazini M, Ostroschi LC, Alencar SM, Massarioli AP, Carvalho Balieiro JC, de Carvalho RA, Favaro-Trindade CS (2017) Characterization of antioxidant and antimicrobial properties of spray-dried extracts from peanut skins. Food Bioprod Process 105:215–223. https://doi.org/10.1016/j.fbp.2017.08.001
14. Evageliou V, Panagopoulou E, Mandala I (2019) Encapsulation of EGCG and esterified EGCG derivatives in double emulsions containing whey protein isolate, bacterial cellulose and salt. Food Chem 281:171–177. https://doi.org/10.1016/j.foodchem.2018.12.105
15. Silva MP, Thomazini M, Holkem AT, Pinho LS, Genovese MI, Fávaro-trindade CS (2019) Production and characterization of solid lipid microparticles loaded with Guaraná (Paullinia Cupana) seed extract. Food Res Int 123:144–152. https://doi.org/10.1016/j.foodres.2019.04.055
16. Fávaro-Trindade CS, de Pinho SC, Rocha GA (2008) Revisão: microencapsulação de ingredientes alimentícios. Brazilian J Food Technol 11:103–112
17. Jafari SM, Assadpoor E, He Y, Bhandari B (2008) Encapsulation efficiency of food flavours and oils during spray drying. Dry Technol 26:816–835. https://doi.org/10.1080/07373930802135972
18. Geranpour M, Assadpour E, Jafari SM (2020) Recent advances in the spray drying encapsulation of essential fatty acids and functional oils. Trends Food Sci Technol 102:71–90. https://doi.org/10.1016/j.tifs.2020.05.028
19. Tulini FL, Souza VB, Thomazini M, Silva MP, Massarioli AP, Alencar SM, Pallone EMJA, Genovese MI, Favaro-Trindade CS (2017) Evaluation of the release profile, stability and antioxidant activity of a proanthocyanidin-rich cinnamon (Cinnamomum Zeylanicum) extract co-encapsulated with α-tocopherol by spray chilling. Food Res Int 95:117–124
20. Okuro PK, Eustáquio de Matos F, Favaro-Trindade CS (2013) Technological challenges for spray chilling encapsulation of functional food ingredients. Food Technol Biotechnol 51:171–182
21. Balassa LL, Fanger GO, Wurzburg OB (1971) Microencapsulation in the food industry. CRC Crit Rev Food Technol 2:245–265. https://doi.org/10.1080/10408397109527123
22. Shahidi F, Han XQ (1993) Encapsulation of food ingredients. Crit Rev Food Sci Nutr 33:501–547. https://doi.org/10.1080/10408399309527645
23. Rosenberg, A.H.F. Fat-soluble vitamin composition and methods of making the same. 1958
24. Arslan-Tontul S, Erbas M, Gorgulu A (2019) The use of probiotic-loaded single- and double-layered microcapsules in cake production. Probiotics Antimicrob Proteins 11:840–

849. https://doi.org/10.1007/s12602-018-9467-y

25. Arslan-Tontul S, Erbas M (2017) Single and double layered microencapsulation of probiotics by spray drying and spray chilling. LWT 81: 160–169. https://doi.org/10.1016/j.lwt.2017.03.060
26. Fadini AL, Alvim ID, Carazzato CA, Paganotti KB d F, Miguel AMR d O, Rodrigues RAF (2021) Microparticles loaded with fish oil: stability studies, food application and sensory evaluation. J Microencapsul 38:365–380. https://doi.org/10.1080/02652048.2021.1948622
27. Fadini AL, Dutra Alvim I, Paganotti KB d F, Bataglia da Silva L, Bonifácio Queiroz M, Miguel AMRd O, Rodrigues RAF (2019) Optimization of the production of double-shell microparticles containing fish oil. Food Sci Technol Int 25:359–369. https://doi.org/10.1177/1082013219825890
28. Fadini AL, Alvim ID, Ribeiro IP, Ruzene LG, da Silva LB, Queiroz MB, Miguel AMR d O, Chaves FCM, Rodrigues RAF (2018) Innovative strategy based on combined microencapsulation technologies for food application and the influence of wall material composition. LWT Food Sci Technol 91:345–352. https://doi.org/10.1016/j.lwt.2018.01.071
29. de Lima PM, Dacanal GC, Pinho LS, de Sá SHG, Thomazini M, Favaro-Trindade CS (2022) Combination of spray-chilling and spray-drying techniques to protect carotenoid-rich extracts from pumpkin (Cucurbita Moschata) byproducts, aiming at the production of a powdered natural food dye. Molecules 27:*7530*. https://doi.org/10.3390/MOLECULES27217530
30. Pinho LS, de Lima PM, de Sá SHG, Chen D, Campanella OH, da Costa Rodrigues CE, Favaro-Trindade CS, de Lima PM, de Sá SHG, Chen D et al (2022) Encapsulation of rich-carotenoids extract from Guaraná; (Paullinia Cupana) byproduct by a combination of spray drying and spray chilling. Food Secur 11. https://doi.org/10.3390/foods11172557
31. de Sá SHG, Chalella Mazzocato M, Saliba ASMC, Alencar SM, Sílvia Favaro-Trindade C (2023) Evaluation of the release, stability and antioxidant activity of Brazilian red Propolis extract encapsulated by spray-drying, spray-chilling and using the combination of both techniques. Food Res Int 164:112423. https://doi.org/10.1016/J.FOODRES.2022.112423
32. Santos PD d F, Batista PS, Torres LCR, Thomazini M, de Alencar SM, Favaro-Trindade CS (2023) Application of spray drying, spray chilling and the combination of both methods to produce Tucumã oil microparticles: characterization, stability, and β-carotene bioaccessibility. Food Res Int 172:113174. https://doi.org/10.1016/J.FOODRES.2023.113174
33. Pinho LS, Lima PM, Fang F, Cooperstone JL, Favaro-Trindade CS, Campanella OH (2024) Effect of extrusion process conditions on extrudates enriched with carotenoids encapsulated by different methods using gum arabic and vegetable fat as carriers. Int J Biol Macromol 267:131200. https://doi.org/10.1016/j.ijbiomac.2024.131200
34. de Oliveira WQ, Angélica Neri Numa I, Alvim ID, Azeredo HMC, Santos LB, Borsoi FT, de Araújo FF, Sawaya ACHF, do Nascimento GC, Clerici MTPS et al (2024) Multilayer microparticles for programmed sequential release of phenolic compounds from Eugenia stipitata: stability and bioavailability. Food Chem 443: 138579. https://doi.org/10.1016/j.foodchem.2024.138579
35. Santos PDd F, Rubio FTV, da Silva MP, Pinho LS, Favaro-Trindade CS (2021) Microencapsulation of carotenoid-rich materials: a review. Food Res Int 147:110571. https://doi.org/10.1016/J.FOODRES.2021.110571

Chapter 7

Encapsulation Using Electrohydrodynamic Processing: Electrospinning and Electrospray

Pedro M. Silva, Vasco D. F. Martins, Lorenzo M. Pastrana, and Miguel A. Cerqueira

Abstract

Electrohydrodynamic processing (EHDP), such as electrospinning and electrospray, is a drying method that produces micro- and nanostructures using food-grade materials. The developed structures can be applied to different areas of the food industry, but they have gained interest in the encapsulation of compounds due to their capacity to form the dried carrier systems in only one step without using high temperatures. This work aims to provide information about the process basics in EHDP technology that deals with controlling its operational parameters to optimize processability and achieve end-product quality and homogeneity as well as the materials used for food applications.

Key words Microparticle, Nanofiber, Polysaccharide, Protein, Food industry, Bioactive compound, Micronutrient

1 Introduction

While electrohydrodynamic processing (EHDP) is regarded as a novel technique, its base concept has been explored for more than a century. In the late nineteenth century and early twentieth century, Rayleigh [1] and Zeleny [2, 3] explored and first described the basic principles of an electrohydrodynamic process. Initial applications focused on the production of fibers in nonwoven threads, came in the 1930s and 1940s via Anton Formhals [4–7]. Further advances were made by Vonnegut and Taylor in the 1950s and 1960s, namely in respect to in-depth studies of the effect of electrical fields on liquid droplets [8–10], leading to the mathematical modulation of the jet formed at the tip of electrified capillaries, now aptly known as "Taylor cone." Until the turn of the century, EHDP remained largely ignored (also due to the challenges of imaging at

Pedro M. Silva and Vasco D. F. Martins contributed equally to this work.

Samantha Cristina de Pinho and Carmen S.F. Trindade (eds.), *Bioactives Encapsulation: Food Applications*, Methods and Protocols in Food Science, https://doi.org/10.1007/978-1-0716-4538-3_7,

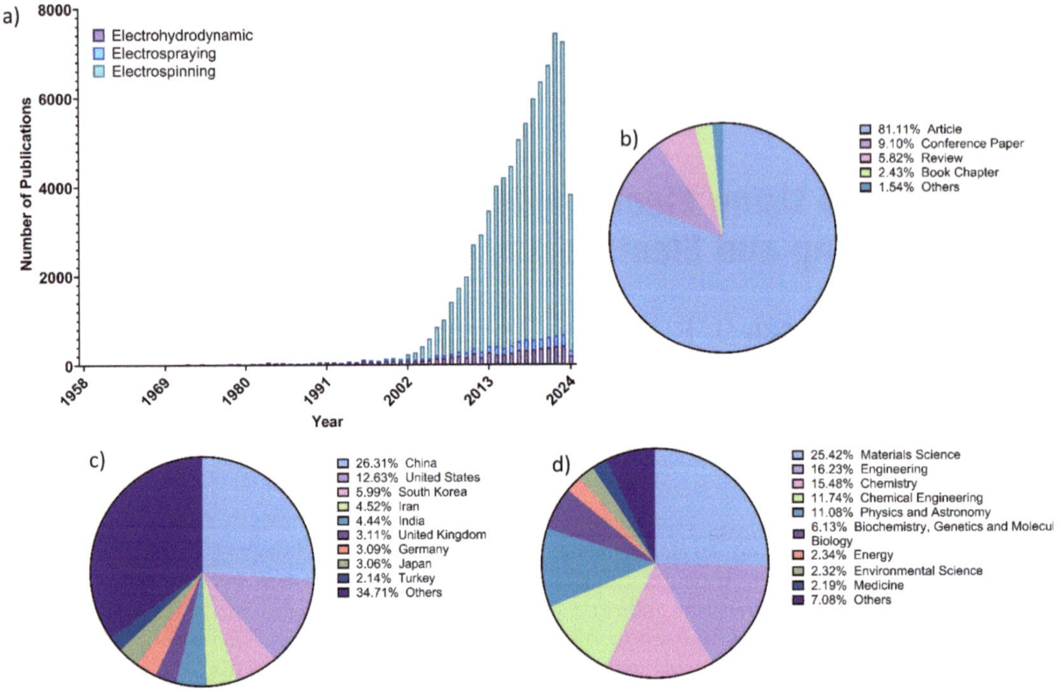

Fig. 1 Scopus data regarding electrohydrodynamic processing (EHDP) publications across time (**a**) and gathered by type of (**b**) publication, (**c**) country, and (**d**) subject area. (Data taken on May 29, 2024. Keywords used: "electrospinning," "electrospun," "electrohydrodynamic," "electrospraying," or "electrosprayed")

micro and nanoscale), when researchers paid renewed attention to its potential for application in the development of ultrafine structures, leading to an increasing interest in EHDP over the past decades (Fig. 1).

Studies on EHDP vary according to objective and aim. Some tackle the process itself, whereas other focus more on the relationship between parameter influence and obtained structures, while other studies focus on the micro-nanostructure optimization and characterization [11–15]. A multitude of polymers can be used for the development of microstructures via EHDP, and these can be applied in a wide range of areas, from biomedicine to food science and to environmental protection, electronics, and others [16–19], which has certainly helped the growth in EHDP publications.

EHDP is a simple and versatile technique to produce micro- and nanostructures that can be operated in two basic modes: electrospinning and electrospraying, which allow for the development of micro-nanofibers and micro- and nanoparticles, respectively. A basic setup for EHDP includes four main components, a feed pump, a syringe with a metal spinneret, a high voltage power supply, and a grounded collector (Fig. 2) [21–23].

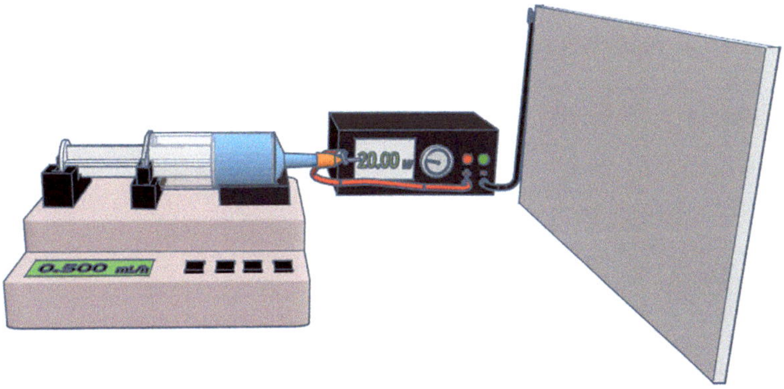

Fig. 2 Schematic illustration of the basic setup of EHDP. (Reprinted from Silva et al. [20] with permission from Elsevier)

When the electrical forces between the spinneret and the collector exceed the surface tension of the polymer, a "Taylor cone" is formed, drawing polymer solution from the spinneret tip, and onto to the collector. This leads to solvent evaporation, producing dried nanofibers (electrospinning) or particles (electrospraying) [15]. Electrospinning relies on the development of a stable steady-state jet, with three main steps of jet initiation, elongation, and solidification, in which bending and whipping instabilities are essential to produce thin, uniform, and dry fibers. Electrospraying, on the other hand, relies on the disruption of the "Taylor cone" via varicose instability (typically due to lower viscosities), transforming ejected jets into droplets instead of fibers. These droplets are further broken up into smaller and stabler droplets via Columbic fission, preventing their agglomeration [23–25].

There is no universal EHDP setup for all applications and polymers, and as such a case-by-case optimization of the parameters involved in electrodynamic processing is required. These parameters are typically divided into three main categories, process parameters, solution properties, and ambient conditions [26, 27].

2 Materials

Several materials can be used in the EHDP, and the capacity to form fibers or particles with food-grade materials is one of the main research topics in the area. Table 1 shows some materials used to produce fibers and particles using EHDP. In some cases, they are studied for the encapsulation of compounds, but some works only evaluate the type of solution, processing conditions, and environmental parameters on the formation of the particles and/or fibers, focusing on multiple applications. It is important to mention that not all materials can produce fibers and particles alone, and

Table 1
Food-grade material used to produce particles and fibers by using EHDP

Category	Source	Materials
Polysaccharide	Plant	Cellulose derivatives (e.g., hydroxypropyl methylcellulose, ethylcellulose); starch; pectin; cyclodextrin; galactomannan; glucomannan; inulin
	Animal/ microorganism	Chitosan; pullulan
	Algae	Alginate; carrageenan; agarose
Protein	Plant	Zein; gluten; soy protein; pea protein; potato protein; bean protein; amaranth protein; algae protein; wheat protein
	Animal	Whey protein; gelatin

normally, they need to be combined with other materials. From the list, the materials that can be highlighted as the ones that can be processed alone and form defined structures are hydroxypropyl methylcellulose (HPMC), ethylcellulose, gelatin, and zein. When considering the material that will be used as wall material for encapsulation, the solvent used is one of the parameters that need to be confirmed since, depending on the compounds that we are willing to encapsulate, different solvents should be used (e.g., water, ethanol, or their mixtures). In this regard, other compounds, such as surfactants, can be used to decrease surface tension and improve the process. Also, the molecular weight and possible chemical interaction with the compound that will be encapsulated should be evaluated. However, when selecting a material, the main factor that should be considered is whether it can be added to the final food product and the acceptable amount according to the region's regulation.

3 Methods

3.1 Solution Properties

Solution parameters such as polymer molecular weight, solution concentration, and viscosity are severely intertwined and play an important role on the structure obtained during EHDP. At lower values of these parameters, particles are a more likely outcome, and when these parameters increase the likelihood of fiber formation increases. This is due to an increase in chain entanglements, which favor the production of fibers [28]. Chain entanglements can be defined as the physical interlocking of polymer chains and are lower in low molecular weight polymers and higher in high molecular weight polymers. The solution entanglement number, $(n_e)_{\mathrm{soln}}$, can be calculated using Eq. 1, as is defined as the ratio between the polymer molecular weight (M_w) and its solution entanglement molecular weight. M_e is the entanglement molecular weight of the undiluted polymers, while $Ø_P$ is the volume fraction [27, 29].

$$(ne)_{soln} = \frac{M_W}{(M_e)_{soln}} = \frac{(\phi_P M_W)}{M_e} \quad (1)$$

Via the changes in entanglements electrospinning and electrospraying zones can be defined, at certain polymer concentrations (c), where viscosity changes abruptly, a point that is defined as the overlap concentration (c^*). When $c << c^*$ (dilute solutions) no chain entanglements exist, and therefore electrospraying typically occurs due to varicose instabilities. Chain overlaps are initiated as concentrations increase ($c \sim c^*$), resulting in a mixture of particles and fibers. As concentration continues to increase ($c > c^*$), the resulting chain entanglements are able to stabilize the electospinning jet, favoring the production of fibers. The spinnability of a solution can thus be estimated via establishing a relationship between particle or fiber formation and the resulting value of $(n_e)_{soln}$. Previous works have demonstrated that electrospraying is favored at $(n_e)_{soln} < 2$ (dilute regime), beaded fibers are favored at $2 < (n_e)_{soln} < 3.5$ (semidilute unentangled regime), and electrospinning is favored at $(n_e)_{soln} > 3.5$ (semidilute entangled regime) [27, 29].

Solvent selection also affects the outcome of EHDP, and it is intrinsically connected with the surface tension of the polymer solution (which is critical for the development of the Taylor cone), volatility, and conductivity. Surface tension and volatility must be within a certain range of values to guarantee solution electrospinabillity. Reducing surface tension is a critical step for EHDP as, for example, the surface tension of water is inappropriate for the formation of a Taylor cone. This decrease can be achieved via mixture of different solvents or through the addition of surfactants. The addition of surfactants to polymer solutions can also lead to changes in conductivity and viscosity [15, 30].

3.2 Process Parameters

The process parameters influence the success of EHDP and the size and final morphology of the obtained structures. These parameters include applied voltage or electric field, feed flow rate of the polymer solution, spinneret morphology (type of spinneret and diameter), collector morphology, and tip-to-collector distance [17, 28, 31, 32]. Table 2 shows their influence on EHDP.

The applied voltage in EHDP is crucial for Taylor cone formation, enabling electrospinning or electrospraying. Besides the impact on particle and fiber morphology, voltage also determines the jet modes, ideally occurring in cone jet mode [45–47]. Different modes include dripping or micro dripping (usually owing to low voltages); spindle (as the applied voltage increases); steady cone jet (appropriate voltage applied); oscillating jet (lateral shifting of the

Table 2
Influence of experimental parameters on EHDP

Category	Parameter	Influence on electrohydrodynamic processing	References
Solution parameters	Polymer molecular weight (M_W), concentration, and viscosity	Low-M_W polymers induce lower solution viscosities, more likely to form particles, due to varicose instability. Higher M_W are more likely to form fibers due to an increase in the solution viscosity and jet stabilization. Increasing polymer concentration leads to a transition from particle formation to beaded fibers to bead-free fibers.	[29, 33–35]
	Solvent	Solvent choice can affect viscosity, surface tension, and conductivity. Solvents should have low but sufficient surface tension, adequate solubility of the polymer, and relatively low volatility, allowing for solvent evaporation and structure deposition on the collector. Extremely high solvent volatility is linked to porous surface morphology and needle clogging. Low volatility impairs solvent evaporation during the process and affects fiber formation.	[17, 36, 37]
	Surface tension	Overcoming surface tension is essential to conduct electrohydrodynamic processing. Low surface tension solutions require less voltage to conduct EHDP. Higher surface tension solutions require more voltage or are not spinnable. Surface tension can be modified through the addition of surfactants. A minimum surface tension is required to avoid process instabilities.	[28, 30, 37, 38]
	Conductivity	Higher conductivity solutions tend to form jets of higher charge producing smoother structures (e.g., smooth surfaces, bead-free fibers) with lower diameters. Excessive conductivity can disrupt the balance between charge droplets and applied voltage, leading to jet instability and morphology changes (e.g., higher diameters, deformed structures).	[34, 36, 37]

Process parameters	Applied voltage	Essential to overcome surface tension. High applied voltages can lead to lower diameters (due to size reduction of the fluid jet) but can also cause particle elongation in electrospraying. Excessive voltage leads to destabilization of the Taylor cone, leading to higher diameters and broad size distribution.	[33, 34, 39, 40]
	Flow rate	Low rates are more adequate in electrospraying processes, producing more spherical particles. Using high rates can lead to the production of beaded fibers or incomplete solvent evaporation, resulting in merged particles and fibers. A steady state is achieved when the feed flow rate equals the flow rate of the ejected solution. Flow rate is the main parameter affecting fiber and bead throughputs.	[33, 34, 39]
	Spinneret morphology	Needle diameter affects the size and stability of the Taylor cone, although reports vary on its influence. Lower diameters can cause needle blockage in electrospinning and are typically more used in electrospraying, where lower viscosity solutions are used.	[33, 39]
	Tip-to-collector distance	As tip-to-collector distance increases, so does the voltage requirement to maintain a stable Taylor cone. Inadequate distances might lead to incomplete solvent evaporation and particle and fiber merger in the collector.	[34, 39]
Ambient parameters	Relative humidity (RH)	RH can increase, decrease, or not affect particle or fiber diameter and morphology (polymer dependent). High RH is known to lead to the development of porous structures. Extremely high RH can make a polymer nonspinnable through the decrease in electrostatic charges at the tip of the spinneret.	[36, 38, 41, 42]
	Temperature	Lower temperatures increase the viscosity of polymers, leading to higher diameter and size distributions. Higher temperatures have the inverse effect. Excessively high temperatures can have a negative effect on the diameter of the structures due to an increase in solvent evaporation.	[43, 44]

Reproduced with permission from Silva et al. [15]

cone jet due to voltage increase); and multijet (as a result of excess voltage, creating multiple cones). The electrical field relies not only on voltage but also on the tip-to-collector distance. One of the fundamental principles of electrohydrodynamic jet printing stands that closer distances reduce voltage requirements and vice versa. Moreover, tip-to-collector distance ensures jet evaporation to obtain uniform structures, preventing droplet splashing and merging [28, 31, 32].

Spinneret morphology's influence on needle electrospinning is debated; nonetheless, its diameter has been recognized to affect Taylor cone stability. In general, smaller diameters favor micro- and nanoparticles when using low viscous solutions but can lead to blockage with higher viscous solutions [33, 39]. Furthermore, feed flow rates have a great influence on structure morphology by affecting transfer rate and jet speed, with higher rates hindering solvent evaporation and causing particle and fiber merging [33, 34]. It has been reported that higher tip-to-collector distances and voltages were more successful in producing stable cones, with larger distances requiring higher voltages [34, 36]. Additionally, low flow rates favor stable cones, while high rates demand higher voltages to counteract solvent evaporation [32–34]. These interconnected effects highlight the importance of balancing process parameters for optimal EHDP outcomes.

3.3 Environmental Parameters

Besides the influence of the solution and process parameters, the environmental parameters during EHDP also play a role in its outcome. For example, temperature changes can lead to modifications of the viscoelastic properties of the polymer solutions, as well as the evaporation rate of the chosen solvents. On the other hand, variations in relative humidity can have an impact on the outcome of the electrohydrodynamic process due to its influence on the development of the electrostatic field [28, 43, 44]. Reports in literature vary, regarding the influence of the ambient conditions. For example, relative humidity has been reported to increase, decrease, or not influence the size or morphology of particles and fibers, depending on the polymer being used. Nonetheless, high values of relative humidity have also been reported to lead to the formation of porous structures, and in some cases, extreme humidity conditions can render a polymer nonspinnable [36, 38, 41, 42].

Temperature, on the other hand, can impact the viscosity of polymers, which in turn can cause alterations in diameter and size distribution, with higher temperatures seemingly improving particle and fiber homogeneity. Nonetheless, extremely high temperatures have also been reported to negatively influence the outcome of EHDP due to increases in the solvent evaporation rate [43, 44].

As such it is clear the influence and impact of the solution, process, and ambient properties in the outcome of EHDP. Controlling these parameters will allow for better homogeneity and tailored properties of the intended micro-nanostructures (e.g., size and surface porosity). Some of these properties are highly interconnected among themselves, so Table 2 should be taken as a starting point not a be-all and end-all of the influence of these parameters on structure morphology and properties. An in-depth study should be conducted for each polymer/solvent combo to properly understand the interactions of these parameters as per the case of HPMC, found in the next section.

3.4 Setup Considerations

EHDP has progressed from basic setups to new and more complex configurations, fostering the production of polymeric micro- and nanostructures with enhanced functionalities and overcoming process drawbacks such as low production yields. The main setup configurations, respective operation modes, and considerations are listed below, as well as the schematic designs of some of them are represented in Fig. 3 and are described as follows:

1. Uniaxial EHDP, employing a single-fluid spinneret, is the most used configuration. It facilitates the formation of various polymeric structures, from single fibers or particles to complex

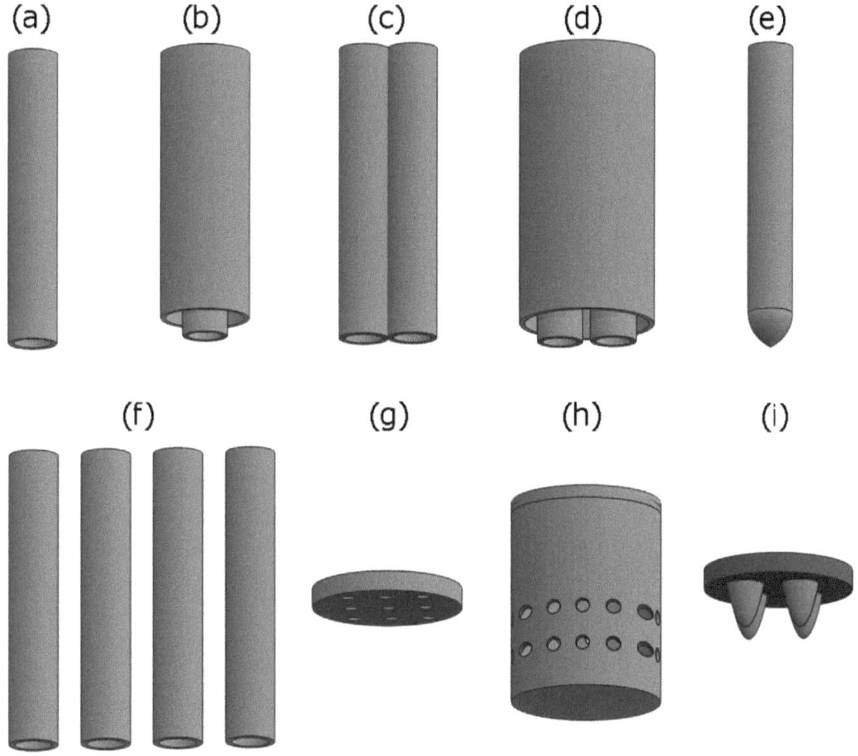

Fig. 3 Schematic illustration of some of the different spinneret configurations available for EHDP: (**a**) uniaxial; (**b**) coaxial; (**c**) side-by-side; (**d**) multifluidic; (**e**) electrohydrodynamic jet printing; (**f**) multineedle; (**g**) needleless; (**h**) centrifugal; and (**i**) multipin. (Reproduced from Silva et al. [15] under Creative Commons CC BY license)

compositions. By employing emulsions or dispersions, diverse structures like ultrathin coatings or composite fibers embedded with micro- or nanoparticles can be achieved [48–58].

2. Multiaxial EHDP introduces smaller needles within larger ones, allowing for more complex structures and enhanced functionality, such as improved protection of encapsulated compounds and controlled release. Coaxial spinnerets are commonly used to form core/shell structures, though optimization is required due to solvent and polymer miscibility issues [59–62].
3. Compound-fluidic configurations involve grouping needles inside a larger one, enabling independent loading of multiple compounds. This way, different core fluids can be fed to the metal capillaries at suitable flow rates, while a viscous shell fluid can be controlled and pumped out from the outer needle. This setup has been exploited to create microcapsules with individual compartments, holding promise for applications like multicomponent drug delivery or microreactors [63–66].
4. Gas-assisted electrospinning/electroblowing: Integrates pressurized gas with high voltages to enhance scalability and throughput [67–69].
5. Side-by-side EHDP places needles in parallel, resulting in the merger of the nanostructures at the tip of the spinneret. It can be used with one or multiple heads and other configurations, allowing for the production of Janus particles and fibers with intrinsic properties of multiple polymers and avoiding solvent miscibility issues. While offering advantages, optimization remains challenging due to fluid dynamics complexities [70–75].
6. Conjugate EHDP uses two spinnable solutions individually added into separate syringes, and then transferred to two processing heads (twin-head), which are oppositely placed and charged, and a collector centered between (above or below) them. The merger of structures occurs when the polymer solutions are in flight to the collector, enabling the production of Janus-structured materials and well-aligned fibers [76–79].
7. Simultaneous electrospinning/electrospraying allows for the simultaneous production of ultrathin fibers and particles. Usually, particles are placed on the surface of fiber mats in a one-step process, offering versatility in creating complex structures [80–82].
8. Electro-netting generates 2D fiber networks in a one-step process, making use of a typical uniaxial setup applying very high voltages [83, 84].

9. Electrohydrodynamic (EHD) jet printing also uses the electrospinning setup, replacing needles with solid probes and reducing the tip-to-collector distance and voltage applied for improved fiber deposition control. This allows creating printing patterns, especially adequate for electronics and sensors. Moreover, it can operate in continuous near-field electrospinning (NFES) or dot-based drop-on-demand (DOD) EHD printing modes [85–88].
10. Multineedle EHDP enhances productivity by applying multiple needles to increase the number of Taylor cones. Studies show that a circular distribution provides more stable cone jets and lower size distribution when compared to a linear distribution, although the production yield is similar [89–91].
11. Needle-less EHDP allows to develop structures with a variety of collectors and non-needle spinnerets with different geometries (e.g., wire, cylinders, discs, among others). It allows an increase in the production yield similar to the multineedle mode; however, due to process limitations, higher voltages are typically needed [92–95].
12. Centrifugal electrospinning combines electrostatic and centrifugal forces to lower voltage requirements and increase production rate. A spinneret with multiple needles (or an array of holes) stores the polymer solution in a rotating reservoir, surrounded by grounded collectors [31, 96, 97].
13. Multipin electrospinning uses solid profiled pins instead of needles, addressing technical issues and limitations in terms of clogging and enhancing Taylor cone formation and stabilization [98].
14. Melt electrospinning, an alternative to solution-based methods, employs heating systems to process thermoplastic materials without solvents, thus increasing throughput. This approach has evolved into complex designs such as melt centrifugal or needle-less melt electrospinning, offering alternatives for various applications [99–102].

In conclusion, EHDP setups continue to evolve, offering diverse configurations and methods to address process limitations and enhance the production of micro- and nanostructures tailored for a wide range of applications.

4 Troubleshooting (Table 3)

Table 3
Common operating problems when using EDHP, their possible cause and suggested solutions

Problem	Possible cause	Suggested solutions
Sparking or electrical discharges	No fluid flow or flows too slowly	Wait for the solution to exit the tip if the flow is temporarily interrupted or increase the feed flow rate
	Voltage is too high	Decrease voltage Increase tip-to-collector distance
	Grounded parts (camera, light source) are too close to the nozzle	Move the grounded parts away from the charged nozzle
	Insulating elements are dirty	Clean insulating elements with an appropriate solvent
A stable Taylor cone cannot be established	No voltage	Check that the HV cable is properly connected to the nozzle Check that liquid or dirty residues have been properly removed Check all safety measures are properly enabled
	Taylor cone does not stabilize and pulsates periodically	Adjust flow rate and voltage
	Dripping or spindle mode	Increase voltage
	Oscillating or multijet mode	Decrease voltage
	Solution with high surface tension	Increase voltage Add appropriate surfactant to the formulation
The solution is not flowing or flows too slowly	Wrong syringe diameter	Verify that the introduced diameter corresponds to the diameter of the syringe in use
	Nozzle or feeding tubes are clogged	Clean nozzle or feeding tubes with appropriate solvent or in an ultrasonic bath
	Solution is too viscous	Reduce the viscosity of the liquid Use a larger diameter spinneret
	Flow rate is too low	Increase pump speed
	High solvent volatility	Choose a different solvent or mix of solvents Choose a different ratio of solvents Try a coaxial setup Increase relative humidity Decrease temperature
Nonhomogeneous structures	Flow rate is too high	Decrease pump speed
	Solution concentration is too low or too high	Adjust solution concentration
	Voltage is too high	Decrease voltage Increase tip-to-collector distance
	Solution with high conductivity	Decrease pump speed
Particles cannot be obtained	High chain entanglements	Use lower M_w polymers or lower concentrations and viscosities
	Flow rate is too high	Decrease pump speed
Bead-free fibers cannot be obtained	Low chain entanglements	Use higher M_w polymers or higher concentrations and viscosities
	Solution with low conductivity	Add appropriate salts to the solution

(continued)

Table 3 (continued)

Problem	Possible cause	Suggested solutions
Particle or fiber merge in the collector	Incomplete solvent evaporation	Increase tip-to-collector distance Increase voltage Decrease pump speed Increase temperature Decrease relative humidity

5 The HPMC Example

There are several examples in the literature of polysaccharides and proteins or more complex structures such as emulsions that have been processed using EHDP, but only a few materials can form particles and fibers by changing their properties and processing conditions. This is related to the possibility of varying the solvent used for the same polymer and the capacity of obtaining polymers where their chemical characteristics can be changed, e.g., the same type of polymer with different molecular weights (M_W). An example is HPMC, which is soluble in water but partially soluble in ethanolic solutions, and can be found with different M_W. HPMC has been widely used in the food industry as a thickening, dispersion, stabilizing, and emulsifying agent for different food applications (e.g., bakery applications). In Silva et al. [27], HPMC of different M_W was adjusted to form fibers and particles according to the concentration used. In this case, and as in most polymers'-case, higher M_W and polymer concentrations result in higher viscosity, affecting the production of fibers or particles (Fig. 4). This influence can be explained by the increase in chain entanglements for higher M_W polymers and at higher concentrations. The higher chain entanglements help stabilize the electrospinning jet, allowing for solvent evaporation and fiber formation. However, when reducing the M_W and concentration, particles are formed.

Later, Silva et al. [26] used the defined conditions to encapsulate melanoidins (up to 2.5 mg/mL) in both particles and fibers. Then, in 2024, they also showed that resveratrol could be encapsulated in the particles in concentrations up to 0.75 mg/mL. Using similar conditions, Guía-García [19] optimized the electrospray conditions to produce microparticles loaded with plant extracts (up to 1 mg/mL). We also know that it is possible to produce loaded particles with HPMC only using water as a solvent, but in that case, a surfactant is needed to decrease the solution's surface tension. Interestingly, the use of solution with different pHs can also be used with HPMC and if required a pH solution of 11 can be used, resulting in round particles (results not published). With these examples, we want to show that food-grade

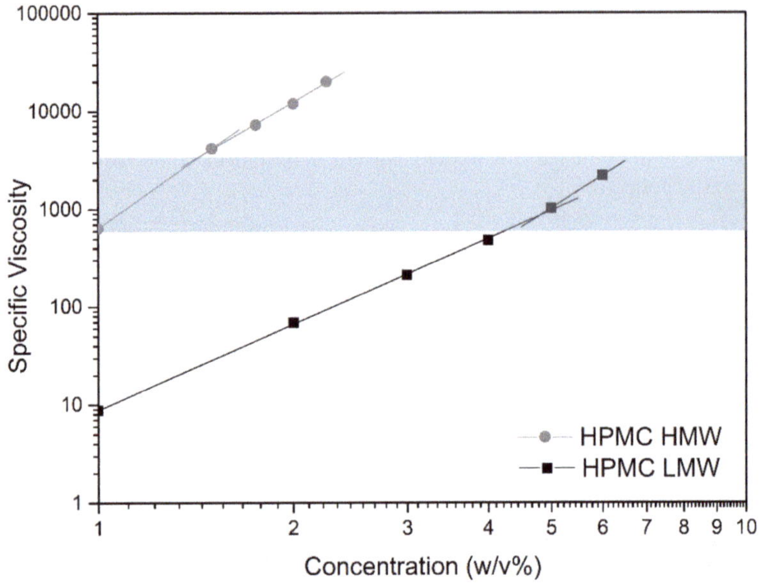

Fig. 4 Specific viscosity for different concentrations of hydroxypropyl methylcellulose (HPMC) low molecular weight (LMW) ($M_W = 90$ kDa) and HPMC high molecular weight (HMW) ($M_W = 746$ kDa). The blue zone represents the conditions where fiber + beads are obtained. Reprinted from Silva et al. [27], with permission from Elsevier

polymers have great potential for developing structures that allow us to encapsulate different compounds, and the production of particles and fibers can be adjusted with processing conditions.

When looking for the encapsulation of a compound, the suggestion is to load the solution with the compound of interest (vitamin, mineral, bioactive, pre- or probiotics) between 5% and 10% in weight of the polymer used and check if the final morphology is maintained, and if possible, increase this concentration until a point where no capsules or fibers are produced. Some adjustments may be needed to optimize the process (e.g., changing voltage, flow rate, etc.).

Acknowledgments

Vasco D.F. Martins is the recipient of a fellowship (SFRH/BD/148504/2019) from the Portuguese Foundation for Science and Technology (FCT). The authors acknowledge the financial support of the project "Agenda VIIAFOOD: Platform for the Agri-Food valorisation, industrialization and commercial innovation," with the reference n.° CC644929456-00000040, co-funded by Component C5— Capitalisation and Business Innovation under the Portuguese Resilience and Recovery Plan, through the NextGenerationEU Fund.

References

1. Rayleigh L (1882) XX. On the equilibrium of liquid conducting masses charged with electricity. Philos Mag Ser 14(87):184–186. https://doi.org/10.1080/14786448208628425
2. Zeleny J (1914) The electrical discharge from liquid points, and a hydrostatic method of measuring the electric intensity at their surfaces. Phys Rev 3(2):69–91. https://doi.org/10.1103/PhysRev.3.69
3. Zeleny J (1917) Instability of electrified liquid surfaces. Phys Rev 10(1):1–6. https://doi.org/10.1103/PhysRev.10.1
4. Formhals A (1934) Process and apparatus for preparing artificial threads. US Patent 1975594, Dec 1934
5. Formhals A (1939) Method and apparatus for spinning. US Patent 2160962, July 1939
6. Formhals A (1940) Artificial thread and method of producing same. US Patent 2187306A, July 1940
7. Formhals A (1943) Production of artificial fibers from fiber forming liquids. United States Patent 2323025
8. Vonnegut B, Neubauer RL (1952) Production of monodisperse liquid particles by electrical atomization. J Colloid Sci 7(6): 616–622. https://doi.org/10.1016/0095-8522(52)90043-3
9. Taylor G (1964) Disintegration of water drops in an electric field. Proc R Soc Lond A Math Phys Sci 280(1382):383–397. https://doi.org/10.1098/rspa.1966.0227
10. Taylor G (1969) Electrically driven jets. Proc R Soc Lond A Math Phys Sci 313(1515): 453–475. https://doi.org/10.1098/rspa.1969.0205
11. Kumar TSM, Kumar KS, Rajini N, Siengchin S, Ayrilmis N, Rajulu AV (2019) A comprehensive review of electrospun nanofibers: food and packaging perspective. Compos Part B Eng 175:107074. https://doi.org/10.1016/j.compositesb.2019.107074
12. Rodrigues RM et al (2020) Electrosprayed whey protein-based nanocapsules for β-carotene encapsulation. Food Chem 314: 126157. https://doi.org/10.1016/j.foodchem.2019.126157
13. Jatoi AW, Ogasawara H, Kim IS, Ni QQ (2020) Cellulose acetate/multi-wall carbon nanotube/Ag nanofiber composite for antibacterial applications. Mater Sci Eng C 110: 110679. https://doi.org/10.1016/j.msec.2020.110679
14. Martins VDF, Cerqueira MA, Fuciños P, Garrido-Maestu A, Curto JMR, Pastrana LM (2018) Active bi-layer cellulose-based films: development and characterization. Cellulose 25(11):6361–6375. https://doi.org/10.1007/s10570-018-2021-y
15. Silva PM, Torres-Giner S, Vicente AA, Cerqueira MA (2022) Management of operational parameters and novel spinneret configurations for the electrohydrodynamic processing of functional polymers. Macromol Mater Eng 307:2100858. https://doi.org/10.1002/mame.202100858
16. Silva PM et al (2024) Resveratrol-loaded octenyl succinic anhydride modified starch emulsions and hydroxypropyl methylcellulose (HPMC) microparticles: cytotoxicity and antioxidant bioactivity assessment after in vitro digestion. Int J Biol Macromol 259: 129288. https://doi.org/10.1016/j.ijbiomac.2024.129288
17. Liao Y, Loh CH, Tian M, Wang R, Fane AG (2018) Progress in electrospun polymeric nanofibrous membranes for water treatment: fabrication, modification and applications. Prog Polym Sci 77:69–94. https://doi.org/10.1016/j.progpolymsci.2017.10.003
18. Li L, Peng S, Lee JKY, Ji D, Srinivasan M, Ramakrishna S (2017) Electrospun hollow nanofibers for advanced secondary batteries. Nano Energy 39:111–139. https://doi.org/10.1016/j.nanoen.2017.06.050
19. Guía-García JL, Charles-Rodríguez AV, Silva P, Robledo-Olivo A, Cerqueira MA, Flores-López ML (2023) Electrosprayed hydroxypropyl methylcellulose microcapsules containing Rhus microphylla fruit extracts and their application in strawberry (Fragaria × ananassa) preservation. LWT 184:115048. https://doi.org/10.1016/j.lwt.2023.115048
20. Silva PM, Torres-Giner S, Vicente AA, Cerqueira MA (2021) Electrohydrodynamic processing for the production of zein-based microstructures and nanostructures. Curr Opin Colloid Interface Sci 56: 101504 https://doi.org/10.1016/j.cocis.2021.101504
21. García-Moreno PJ, Mendes AC, Jacobsen C, Chronakis IS (2018) Biopolymers for the nanomicroencapsulation of bioactive ingredients by electrohydrodynamic processing. In: Polymers for food applications. Springer, pp 447–479. https://doi.org/10.1007/978-3-319-94625-2_17

22. Marques AM, Azevedo MA, Teixeira JA, Pastrana LM, Gonçalves C, Cerqueira MA (2019) Engineered nanostructures for enrichment and fortification of foods. In: Food applications of nanotechnology. CRC Press, pp 61–97. https://doi.org/10.1201/9780429297038-4
23. Wang J, Jansen JA, Yang F (2019) Electrospraying: possibilities and challenges of engineering carriers for biomedical applications – a mini review. Front Chem 7, no. APR:1–9. https://doi.org/10.3389/fchem.2019.00258
24. Fernández de la Mora J (2007) The fluid dynamics of Taylor cones. Annu Rev Fluid Mech 39(1):217–243. https://doi.org/10.1146/annurev.fluid.39.050905.110159
25. Gañán-Calvo AM, López-Herrera JM, Herrada MA, Ramos A, Montanero JM (2018) Review on the physics of electrospray: from electrokinetics to the operating conditions of single and coaxial Taylor cone-jets, and AC electrospray. J Aerosol Sci 125(May):32–56. https://doi.org/10.1016/j.jaerosci.2018.05.002
26. Silva PM et al (2022) Hydroxypropyl methylcellulose-based micro- and nanostructures for encapsulation of melanoidins: effect of electrohydrodynamic processing variables on morphological and physicochemical properties. Int J Biol Macromol 202:453–467. https://doi.org/10.1016/j.ijbiomac.2022.01.019
27. Silva PM et al (2021) Food-grade hydroxypropyl methylcellulose-based formulations for electrohydrodynamic processing: Part I—Role of solution parameters on fibre and particle production. Food Hydrocoll 118:106761. https://doi.org/10.1016/j.foodhyd.2021.106761
28. Alehosseini A, Ghorani B, Sarabi-Jamab M, Tucker N (2018) Principles of electrospraying: a new approach in protection of bioactive compounds in foods. Crit Rev Food Sci Nutr 58(14):2346–2363. https://doi.org/10.1080/10408398.2017.1323723
29. Shenoy SL, Bates WD, Frisch HL, Wnek GE (2005) Role of chain entanglements on fiber formation during electrospinning of polymer solutions: good solvent, non-specific polymer-polymer interaction limit. Polymer (Guildf) 46(10):3372–3384. https://doi.org/10.1016/j.polymer.2005.03.011
30. Fang W, Yang S, Yuan TQ, Charlton A, Sun RC (2017) Effects of various surfactants on alkali lignin electrospinning ability and spun fibers. Ind Eng Chem Res 56(34):9551–9559. https://doi.org/10.1021/acs.iecr.7b02494
31. SalehHudin HS, Mohamad EN, Mahadi WNL, Muhammad Afifi A (2018) Multiple-jet electrospinning methods for nanofiber processing: a review. Mater Manuf Process 33(5):479–498. https://doi.org/10.1080/10426914.2017.1388523
32. Haider A, Haider S, Kang IK (2018) A comprehensive review summarizing the effect of electrospinning parameters and potential applications of nanofibers in biomedical and biotechnology. Arab J Chem 11(8):1165–1188. https://doi.org/10.1016/j.arabjc.2015.11.015
33. Correia DM et al (2014) Electrosprayed poly (vinylidene fluoride) microparticles for tissue engineering applications. RSC Adv 4(62):33013–33021. https://doi.org/10.1039/c4ra04581e
34. Smeets A, Clasen C, Van den Mooter G (2017) Electrospraying of polymer solutions: study of formulation and process parameters. Eur J Pharm Biopharm 119:114–124. https://doi.org/10.1016/j.ejpb.2017.06.010
35. McKee MG, Wilkes GL, Colby RH, Long TE (2004) Correlations of solution rheology with electrospun fiber formation of linear and branched polyesters. Macromolecules 37(5):1760–1767. https://doi.org/10.1021/ma035689h
36. Topuz F, Satilmis B, Uyar T (2019) Electrospinning of uniform nanofibers of Polymers of Intrinsic Microporosity (PIM-1): the influence of solution conductivity and relative humidity. Polymer (Guildf) 178:121610. https://doi.org/10.1016/j.polymer.2019.121610
37. Zhang S, Campagne C, Salaün F (2019) Influence of solvent selection in the electrospraying process of polycaprolactone. Appl Sci (Switzerland) 9(3). https://doi.org/10.3390/app9030402
38. Yazgan G et al (2015) Tunable release of hydrophilic compounds from hydrophobic nanostructured fibers prepared by emulsion electrospinning. Polymer (Guildf) 66:268–276. https://doi.org/10.1016/j.polymer.2015.04.045
39. Jain E, Scott KM, Zustiak SP, Sell SA (2015) Fabrication of polyethylene glycol-based hydrogel microspheres through electrospraying. Macromol Mater Eng 300(8):823–835. https://doi.org/10.1002/mame.201500058

40. Zhou H et al (2019) The relationships between process parameters and polymeric nanofibers fabricated using a modified coaxial electrospinning. Nano 9(6):843. https://doi.org/10.3390/nano9060843
41. Pelipenko J, Kristl J, Janković B, Baumgartner S, Kocbek P (2013) The impact of relative humidity during electrospinning on the morphology and mechanical properties of nanofibers. Int J Pharm 456(1):125–134. https://doi.org/10.1016/j.ijpharm.2013.07.078
42. Huang X, Gao J, Zheng N, Li W, Xue H, Li RKY (2017) Influence of humidity and polymer additives on the morphology of hierarchically porous microspheres prepared from non-solvent assisted electrospraying. Colloids Surf A Physicochem Eng Asp 517:17–24. https://doi.org/10.1016/j.colsurfa.2017.01.003
43. Wang XX et al (2015) Monodisperse erythrocyte-sized and acid-soluble chitosan microspheres prepared via electrospraying. RSC Adv 5(43):34243–34250. https://doi.org/10.1039/c5ra04726a
44. Yang GZ, Li HP, Yang JH, Wan J, Yu DG (2017) Influence of working temperature on the formation of electrospun polymer nanofibers. Nanoscale Res Lett 12(1):55. https://doi.org/10.1186/s11671-016-1824-8
45. Cloupeau M, Prunet-Foch B (1990) Electrostatic spraying of liquids: main functioning modes. J Electrost 25(2):165–184. https://doi.org/10.1016/0304-3886(90)90025-Q
46. Cloupeau M, Prunet-Foch B (1994) Electrohydrodynamic spraying functioning modes: a critical review. J Aerosol Sci 25(6):1021–1036. https://doi.org/10.1016/0021-8502(94)90199-6
47. Zong H et al (2018) Designing function-oriented artificial nanomaterials and membranes via electrospinning and electrospraying techniques. Mater Sci Eng C 92:1075–1091. https://doi.org/10.1016/j.msec.2017.11.007
48. Javadi A, Solouk A, Haghbin Nazarpak M, Bagheri F (2019) Surface engineering of titanium-based implants using electrospraying and dip coating methods. Mater Sci Eng C 99:620–630. https://doi.org/10.1016/j.msec.2019.01.027
49. Wang H, Li W, Li Z (2018) Preparation of fluorinated PCL porous microspheres and a super-hydrophobic coating on fabrics: via electrospraying. Nanoscale 10(39):18857–18868. https://doi.org/10.1039/c8nr05793a
50. Figueroa-Lopez KJ et al (2020) Development of active barrier multilayer films based on electrospun antimicrobial hot-tack food waste derived poly(3-hydroxybutyrate-co-3-hydroxyvalerate) and cellulose nanocrystal interlayers. Nano 10(12):1–24. https://doi.org/10.3390/nano10122356
51. Melendez-Rodriguez B et al (2021) High-oxygen-barrier multilayer films based on polyhydroxyalkanoates and cellulose nanocrystals. Nano 11(6):1443. https://doi.org/10.3390/nano11061443
52. Balik BA, Argin S, Lagaron JM, Torres-Giner S (2019) Preparation and characterization of electrospun pectin-based films and their application in sustainable aroma barrier multilayer packaging. Appl Sci (Switzerland) 9(23):5136. https://doi.org/10.3390/app9235136
53. Figueroa-Lopez KJ, Cabedo L, Lagaron JM, Torres-Giner S (2020) Development of electrospun poly(3-hydroxybutyrate-co-3-hydroxyvalerate) monolayers containing eugenol and their application in multilayer antimicrobial food packaging. Front Nutr 7:140. https://doi.org/10.3389/fnut.2020.00140
54. Quiles-Carrillo L, Montanes N, Lagaron JM, Balart R, Torres-Giner S (2019) Bioactive multilayer polylactide films with controlled release capacity of gallic acid accomplished by incorporating electrospun nanostructured coatings and interlayers. Appl Sci (Switzerland) 9(3). https://doi.org/10.3390/app9030533
55. Hu J, Prabhakaran MP, Ding X, Ramakrishna S (2015) Emulsion electrospinning of polycaprolactone: influence of surfactant type towards the scaffold properties. J Biomater Sci Polym Ed 26(1):57–75. https://doi.org/10.1080/09205063.2014.982241
56. Torres-Giner S, Martinez-Abad A, Lagaron JM (2014) Zein-based ultrathin fibers containing ceramic nanofillers obtained by electrospinning. II. Mechanical properties, gas barrier, and sustained release capacity of biocide thymol in multilayer polylactide films. J Appl Polym Sci 131(18):9270–9276. https://doi.org/10.1002/app.40768
57. Pardo-Figuerez M, López-Córdoba A, Torres-Giner S, Lagaron JM (2018) Superhydrophobic bio-coating made by co-continuous electrospinning and electrospraying polyethylene terephthalate films proposed as easy emptying transparent food packaging. Coatings 8(10):364. https://doi.org/10.3390/coatings8100364

58. Lasprilla-Botero J, Torres-Giner S, Pardo-Figuerez M, Álvarez-Láinez M, Lagaron JM (2018) Superhydrophobic bilayer coating based on annealed electrospun ultrathin poly (ε-caprolactone) fibers and electrosprayed nanostructured silica microparticles for easy emptying packaging applications. Coatings 8(5):173. https://doi.org/10.3390/coatings8050173
59. Yu DG, Zheng XL, Yang Y, Li XY, Williams GR, Zhao M (2019) Immediate release of helicid from nanoparticles produced by modified coaxial electrospraying. Appl Surf Sci 473:148–155. https://doi.org/10.1016/j.apsusc.2018.12.147
60. Tian J et al (2020) Comparison of different electrospinning technologies for the production of arrays with multifunctional properties: fluorescence, conduction and magnetism. J Phys D Appl Phys 53(15):155301. https://doi.org/10.1088/1361-6463/ab69ac
61. Huang RM, Feng K, Li SF, Zong MH, Wu H, Han SY (2021) Enhanced survival of probiotics in the electrosprayed microcapsule by addition of fish oil. *J Food Eng* 307:110650. https://doi.org/10.1016/j.jfoodeng.2021.110650
62. Abdelhakim HE, Coupe A, Tuleu C, Edirisinghe M, Craig DQM (2021) Utilising co-axial electrospinning as a taste-masking technology for paediatric drug delivery. Pharmaceutics 13(10). https://doi.org/10.3390/pharmaceutics13101665
63. Chen H, Zhao Y, Jiang L (2009) Compound-fluidic electrospray: an efficient method for the fabrication of microcapsules with multi-compartment structure. Chin Sci Bull 54(18): 3147–3153. https://doi.org/10.1007/s11434-009-0222-3
64. Chen H, Zhao Y, Song Y, Jiang L (2008) One-step multicomponent encapsulation by compound-fluidic electrospray. J Am Chem Soc 130(25):7800–7801. https://doi.org/10.1021/ja801803x
65. Li Y et al (2013) Colorimetric sensor strips for lead (II) assay utilizing nanogold probes immobilized polyamide-6/nitrocellulose nano-fibers/nets. Biosens Bioelectron 48: 244–250. https://doi.org/10.1016/j.bios.2013.03.085
66. Zhang J et al (2020) Fabrication of flexible TiO2-SiO2 composite nanofibers with variable structure as efficient adsorbent. Ceram Int 46(3):3543–3549. https://doi.org/10.1016/j.ceramint.2019.10.071
67. Busolo MA, Torres-Giner S, Prieto C, Lagaron JM (2019) Electrospraying assisted by pressurized gas as an innovative high-throughput process for the microencapsulation and stabilization of docosahexaenoic acid-enriched fish oil in zein prolamine. Innov Food Sci Emerg Technol 51:12–19. https://doi.org/10.1016/j.ifset.2018.04.007
68. Shirazi MMA, Bazgir S, Meshkani F (2020) A novel dual-layer, gas-assisted electrospun, nanofibrous SAN4-HIPS membrane for industrial textile wastewater treatment by direct contact membrane distillation (DCMD). J Water Process Eng 36:101315. https://doi.org/10.1016/j.jwpe.2020.101315
69. Ahmed J, Matharu RK, Shams T, Illangakoon UE, Edirisinghe M (2018) A comparison of electric-field-driven and pressure-driven fiber generation methods for drug delivery. Macromol Mater Eng 303(5):1–11. https://doi.org/10.1002/mame.201700577
70. Yang GZ, Li JJ, Yu DG, He MF, Yang JH, Williams GR (2017) Nanosized sustained-release drug depots fabricated using modified tri-axial electrospinning. Acta Biomater 53: 233–241. https://doi.org/10.1016/j.actbio.2017.01.069
71. Chen G, Xu Y, Yu DG, Zhang DF, Chatterton NP, White KN (2015) Structure-tunable Janus fibers fabricated using spinnerets with varying port angles. Chem Commun 51(22): 4623–4626. https://doi.org/10.1039/c5cc00378d
72. Sanchez-Vazquez B, Amaral AJR, Yu DG, Pasparakis G, Williams GR (2017) Electrosprayed Janus particles for combined photo-chemotherapy. AAPS PharmSciTech 18(5): 1460–1468. https://doi.org/10.1208/s12249-016-0638-4
73. Yu DG, Li JJ, Zhang M, Williams GR (2017) High-quality Janus nanofibers prepared using three-fluid electrospinning. Chem Commun 53(33):4542–4545. https://doi.org/10.1039/c7cc01661a
74. Ji SM, Tiwari AP, Oh HJ, Kim HY (2021) ZnO/Ag nanoparticles incorporated multifunctional parallel side by side nanofibers for air filtration with enhanced removing organic contaminants and antibacterial properties. Colloids Surf A Physicochem Eng Asp 621: 126564. https://doi.org/10.1016/j.colsurfa.2021.126564
75. Zhang Z et al (2021) Janus rod-like micromotors to promote the tumor accumulation and cell internalization of therapeutic agents. Chem Eng J 404:127073. https://doi.org/10.1016/j.cej.2020.127073
76. Mou F, Chen C, Guan J, Chen DR, Jing H (2013) Oppositely charged twin-head

electrospray: a general strategy for building Janus particles with controlled structures. Nanoscale 5(5):2055–2064. https://doi.org/10.1039/c2nr33523a
77. Tian J et al (2019) High pairing rate Janus-structured microfibers and array: high-efficiency conjugate electrospinning fabrication, structure analysis and co-instantaneous multifunctionality of anisotropic conduction, magnetism and enhanced red fluorescence. RSC Adv 9(19):10679–10692. https://doi.org/10.1039/c9ra01147a
78. Kuang H et al (2020) Construction and performance evaluation of Hep/silk-PLCL composite nanofiber small-caliber artificial blood vessel graft. Biomaterials 259:120288. https://doi.org/10.1016/j.biomaterials.2020.120288
79. Jin D, Wu S, Kuang H, Zhang P, Yin M (2021) Preliminary application of a cell-free mono-layered vascular scaffold in a rabbit model. Mater Des 198:109301. https://doi.org/10.1016/j.matdes.2020.109301
80. Korina E, Stoilova O, Manolova N, Rashkov I (2018) Polymer fibers with magnetic core decorated with titanium dioxide prospective for photocatalytic water treatment. J Environ Chem Eng 6(2):2075–2084. https://doi.org/10.1016/j.jece.2018.03.016
81. Devolder RJ, Bae H, Lee J, Kong H (2011) Directed blood vessel growth using an angiogenic microfiber/microparticle composite patch. Adv Mater 23(28):3139–3143. https://doi.org/10.1002/adma.201100823
82. Spasova M, Stoilova O, Manolova N, Rashkov I, Naydenov M (2020) Electrospun eco-friendly materials based on poly (3-hydroxybutyrate) (PHB) and TiO2 with antifungal activity prospective for esca treatment. Polymers (Basel) 12(6):1384. https://doi.org/10.3390/polym12061384
83. Zhang S, Liu H, Tang N, Ge J, Yu J, Ding B (2019) Direct electronetting of high-performance membranes based on self-assembled 2D nanoarchitectured networks. Nat Commun 10(1):1–11. https://doi.org/10.1038/s41467-019-09444-y
84. Liu H, Liu L, Yu J, Yin X, Ding B (2020) High-efficiency and super-breathable air filters based on biomimetic ultrathin nanofiber networks. Compos Commun 22:100493. https://doi.org/10.1016/j.coco.2020.100493
85. Sun D, Chang C, Li S, Lin L (2006) Near-field electrospinning. Nano Lett 6(4):839–842. https://doi.org/10.1021/nl0602701
86. George D et al (2020) Fabrication of patterned graphitized carbon wires using low voltage near-field electrospinning, pyrolysis, electrodeposition, and chemical vapor deposition. Microsyst Nanoeng 6(1):1–12. https://doi.org/10.1038/s41378-019-0117-7
87. Liashenko I, Rosell-Llompart J, Cabot A (2020) Ultrafast 3D printing with submicrometer features using electrostatic jet deflection. Nat Commun 11(1):1–9. https://doi.org/10.1038/s41467-020-14557-w
88. Liu S, Sun L, Zhang H, Hu Q, Wang Y, Ramalingam M (2021) High-resolution combinatorial 3D printing of gelatin-based biomimetic triple-layered conduits for nerve tissue engineering. Int J Biol Macromol 166:1280–1291. https://doi.org/10.1016/j.ijbiomac.2020.11.010
89. Parhizkar M et al (2017) Performance of novel high throughput multi electrospray systems for forming of polymeric micro/nanoparticles. Mater Des 126:73–84. https://doi.org/10.1016/j.matdes.2017.04.029
90. Xie S, Zeng Y (2012) Effects of electric field on multineedle electrospinning: experiment and simulation study. Ind Eng Chem Res 51(14):5336–5345. https://doi.org/10.1021/ie2020763
91. Yan G, Niu H, Lin T (2018) Needle-less electrospinning. In: Electrospinning: nanofabrication and applications, vol 2. Elsevier Inc, pp 219–247. https://doi.org/10.1016/B978-0-323-51270-1.00007-8
92. Zhang C, Chang MW, Ahmad Z, Hu W, Zhao D, Li JS (2015) Stable single device multi-pore electrospraying of polymeric microparticles via controlled electrostatic interactions. RSC Adv 5(107):87919–87923. https://doi.org/10.1039/c5ra18482g
93. Ryšánek P, Benada O, Tokarský J, Syrový M, Čapková P, Pavlík J (2019) Specific structure, morphology, and properties of polyacrylonitrile (PAN) membranes prepared by needleless electrospinning; Forming hollow fibers. Mater Sci Eng C 105:110151. https://doi.org/10.1016/j.msec.2019.110151
94. Yu W et al (2021) Graphene oxide-silver nanocomposites embedded nanofiber core-spun yarns for durable antibacterial textiles. J Colloid Interface Sci 584:164–173. https://doi.org/10.1016/j.jcis.2020.09.092
95. Wei K et al (2021) Dissymmetric interface design of SnO2/TiO2 side-by-side bi-component nanofibers as photoanodes for dye sensitized solar cells: facilitated electron transport and enhanced carrier separation. J

Colloid Interface Sci 583:24–32. https://doi.org/10.1016/j.jcis.2020.09.017

96. Chang WM, Wang CC, Chen CY (2019) Fabrication of ultra-thin carbon nanofibers by centrifuged-electrospinning for application in high-rate supercapacitors. Electrochim Acta 296:268–275. https://doi.org/10.1016/j.electacta.2018.08.048
97. Huang CY, Chang MH (2021) Fabrication of platinum nanowires by centrifugal electrospinning method for proton exchange membrane fuel cell. Int J Energy Res 45:12968. https://doi.org/10.1002/er.6627
98. Prabu GTV, Dhurai B (2020) A novel profiled multi-pin electrospinning system for nanofiber production and encapsulation of nanoparticles into nanofibers. Sci Rep 10(1):1–11. https://doi.org/10.1038/s41598-020-60752-6
99. Yang W, Li H, Chen X (2019) Melt electrospinning. In: Electrospinning: nanofabrication and applications. Elsevier, pp 339–361. https://doi.org/10.1016/b978-0-323-51270-1.00011-x
100. Ibrahim YS, Hussein EA, Zagho MM, Abdo GG, Elzatahry AA (2019) Melt electrospinning designs for nanofiber fabrication for different applications. Int J Mol Sci 20(10). https://doi.org/10.3390/ijms20102455
101. Nam HC, Park WH (2020) Eco-friendly poly (lactic acid) microbeads for cosmetics via melt electrospraying. Int J Biol Macromol 157:734–742. https://doi.org/10.1016/j.ijbiomac.2019.11.240
102. Buivydiene D, Todea AM, Asbach C, Krugly E, Martuzevicius D, Kliucininkas L (2021) Composite micro/nano fibrous air filter by simultaneous melt and solution electrospinning. J Aerosol Sci 154:105754. https://doi.org/10.1016/j.jaerosci.2021.105754

Chapter 8

Liposome Synthesis by the Hydration of Micronized Sucrose-Coated Proliposomes for the Coencapsulation of Hydrophobic Compounds

Matheus Andrade Chaves and Samantha Cristina de Pinho

Abstract

Hydration of proliposomes presents a cost-effective method for liposome production, offering enhanced scalability and formation of stable vesicles. By using this technique, we can efficiently produce up to 4 g of proliposomes, enabling the production of 100 mL of a liposome dispersion that remains stable for over 14 days. Additionally, this approach affords flexibility, allowing for the formulation of liposomes in either solid or liquid form tailored to specific applications. Here, we demonstrate the encapsulation of two hydrophobic bioactives, curcumin and cholecalciferol, underscoring the efficacy and versatility of proliposome hydration in liposomal delivery systems for food applications.

Key words Nanoencapsulation, Curcumin, Vitamin D_3, Lipid carrier, Cost-effective encapsulation

1 Introduction

In recent years, there has been a notable increase in the utilization of liposomes within the food industry, particularly for targeted dietary supplements [1]. Liposomes present significant advantages, including high encapsulation efficiency, improved solubility of bioactive compounds, and protection against degradation factors such as pH, light, and digestive enzymes [2]. These colloidal structures, as defined by the FDA (Food and Drug Administration), are self-assembled closed vesicles characterized by a phospholipid bilayer structure, varying widely in size from 10 nm to several micrometers, in which both hydrophobic and hydrophilic bioactive molecules can be encapsulated simultaneously [3]. To overcome challenges in liposome production aimed at achieving a long-term shelf life, proliposomes have emerged as a promising solution. These structures are defined as carbohydrate carriers that rapidly

Samantha Cristina de Pinho and Carmen S.F. Trindade (eds.), *Bioactives Encapsulation: Food Applications*, Methods and Protocols in Food Science, https://doi.org/10.1007/978-1-0716-4538-3_8,

transform into liposomal suspensions upon contact with water or biological fluids, providing a straightforward and reproducible technique for large-scale liposome production [4].

Spray-drying and freeze-drying represent the two primary methods for preparing proliposomes. Freeze-drying particularly stands out as the preferred approach for formulating proliposomes containing heat-sensitive payloads due to its ability to avoid high-temperature damage while maintaining optimal thermal efficiency. Nonetheless, the freeze-drying process introduces the risk of ice crystal formation, which can potentially damage the phospholipid bilayer. To address this challenge, the addition of excipients to the proliposome formulation becomes imperative during the dehydration process [5].

In light of these challenges associated with spray-drying and freeze-drying, the production of proliposomes via deposition onto micronized carbohydrates emerges as an alternative solution. Previous studies have highlighted the efficacy of specific nonreducing disaccharides, such as sucrose, trehalose, and sugar alcohol, in preserving the stability of proliposomes [5]. These compounds serve dual roles as lyoprotectants and carriers, establishing hydrogen bonds with the polar head groups of phospholipid bilayers. By displacing water molecules surrounding polar residues, these lyoprotectants effectively prevent the fusion of phospholipid liposome membranes in the dehydrated state [6].

Here, we will describe the production of proliposomes using the micronized sucrose coating (MSC) method. This approach involves initially mixing lipids with a solid material, specifically water-soluble sucrose, resulting in the formation of lipid-coated sucrose particles. Upon hydration, the solid material is expected to dissolve, allowing the lipids to reorganize and form liposomes. The utilization of sucrose as a micronized carrier leads to the formation of more stable liposomes after proliposome hydration compared to vesicles obtained through other methods [7]. Furthermore, sucrose promotes uniformity in vesicle size, particularly when compared to other carbohydrate carriers such as trehalose and lactose [8]. Another advantage of the MSC method is its ability to operate at lower temperatures, making it suitable for incorporating thermolabile bioactive compounds like vitamins [9, 10].

2 Materials

Ensure that all solutions are prepared using reagents of analytical grade. Reagents must be prepared and stored at room temperature unless otherwise instructed. Please comply with all regulations regarding waste disposal when disposing of waste materials. The choice of buffer for liposome production may vary depending on the intended application. For this protocol, we used ultrapure water, obtained by purifying deionized water to achieve a sensitivity of 18 MΩ-cm at 25 °C. It is crucial to emphasize that sodium azide is not included in our reagents but may be present only in the final liposome suspension.

2.1 Cholecalciferol Stock Solution

1. With the assistance of a spatula with a 3 mm scoop, weigh 0.01 g (10 mg) of cholecalciferol powder (>98% purity, Sigma-Aldrich, St. Louis, MO, USA) using an analytical balance and place it in a crucible.
2. Pour the content into a beaker containing 100 mL of ethanol.
3. Ensure thorough mixing by using a magnetic stirrer set at 3200 rpm for 30 min at 25 °C. For this purpose, use a magnetic stir bar according to the magnetic stirrer used, in order to observe the formation of a vortex inside the liquid. After this period, the cholecalciferol stock solution will be ready for use. This solution will contain nearly 400,000 IU of the bioactive compound. The volume of methanol solution to be added during the proliposome production should be proportional to the number of international units of interest.
4. Store the cholecalciferol stock solution at 10 °C protected from light for a maximum of 2 weeks.

2.2 Micronization of Sucrose

1. With the assistance of a spatula with a 20 cm scoop, weigh 20 g of sucrose (for analysis, Synth, Diadema, SP, Brazil) using an analytical balance and place it in a crucible.
2. Transfer the weighed sucrose to a 250 mL plastic laboratory bottle and add 40 g of zirconium balls to the same bottle.
3. Place the plastic laboratory bottle in a lab-scale ball mill and set it to stir constantly at 100 rpm for 10 h at 25 °C.
4. Separate the micronized sucrose from the zirconium balls using a stainless steel granulometric sieve with a compatible mesh <1 mm.
5. Wash the zirconium balls thoroughly and store them in a proper container until future use.
6. Store the micronized sucrose in an amber bottle, properly sealed, in a desiccator containing silica gel at 25 °C and at a relative humidity of 25%.
7. Store the micronized sucrose for a maximum of 3 months.

2.3 Production of Curcumin/Cholecalciferol-co-Loaded Proliposomes

1. Micronized sucrose prepared previously.
2. Cholecalciferol stock solution prepared previously.
3. Curcumin (from *Curcuma longa*, Turmeric, powder, C1386, Sigma-Aldrich, St. Louis, MO, USA).

4. Phospholipon 90H, minimum of 94% total phospholipids, >90% phosphatidylcholine (Lipoid GmbH, Ludwigshafen, Germany).
5. Absolute ethanol (reagent grade).

Apparatus:

6. High-performance dispersing machine (at the researcher's choice, we recommend T25 Ultra-Turrax®, IKA, Staufen, Germany).
7. Peristaltic pump (at the researcher's choice, we recommend Masterflex 7528–30, Cole-Palmer, Vernon Hills, IL, USA).
8. Rotary evaporator.
9. Vacuum pump.
10. Thermostatic bath.
11. 250 mL bottle (Schott).
12. 250 mL tall-form beaker.
13. Silicone hoes (Ø = 1 cm).

2.4 Production of Curcumin/Cholecalciferol-co-Loaded Liposome Dispersions

1. Proliposomes produced previously.
2. Deionized water.
3. Xanthan gum (at the researcher's choice).
4. Guar gum (at the researcher's choice).
5. Sodium benzoate (at the researcher's choice).

Apparatus:

6. High-performance dispersing machine.
7. Magnetic stirrer.

3 Methods

3.1 Production of Curcumin/Cholecalciferol-co-Loaded Proliposomes

Assembling the Equipment

An example of a possible arrangement for the apparatus used for proliposome production is presented in Fig. 1.

1. Connect two silicon hoses to the designated places on the thermostatic bath **(A)** in such a way that the outlet of the bath is connected to the inlet of the condenser **(B)**, and another hose is exiting the condenser and returning to the bath.
2. Connect the middle portion of a silicon hose to the peristaltic pump **(C)** in such a way that one end reaches the bottom of a 250 mL Schott **(D)** and the other end can be attached to the feeding/inlet valve. If the lid of the Schott flask is not a screw

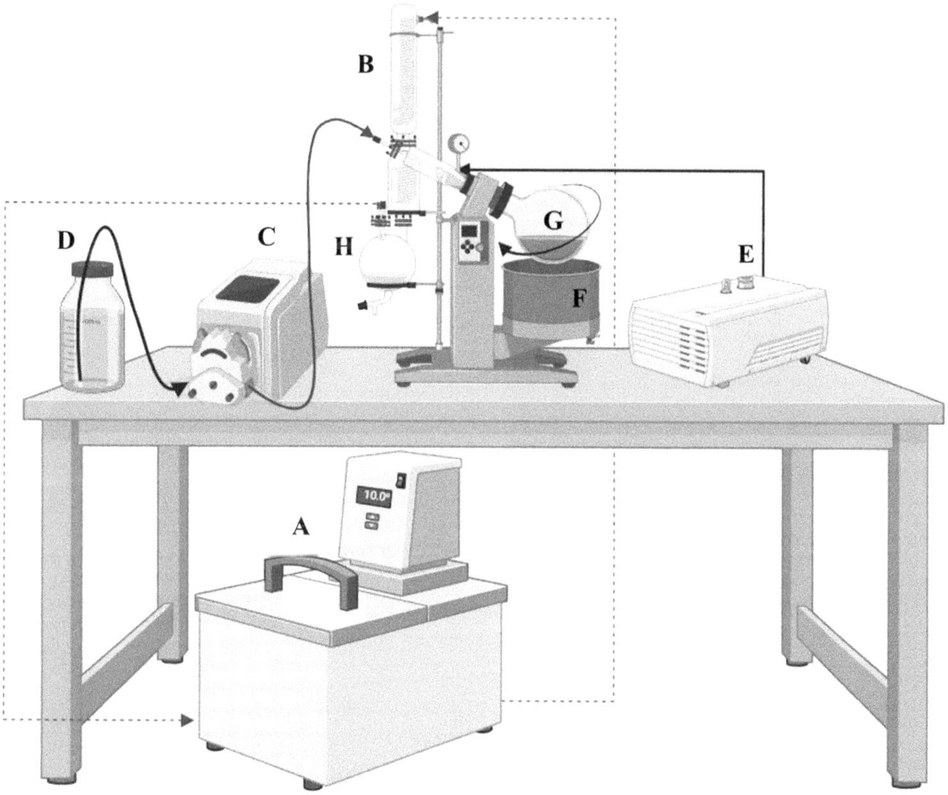

Fig. 1 Example of a schematic representation of equipment used for the production of proliposomes using the micronized sucrose coating method: (**a**) thermostatic bath; (**b**) vertical condenser; (**c**) peristaltic pump; (**d**) 250 mL Schott (which will contain the ethanolic bioactive solution); (**e**) vacuum pump; (**f**) rotary evaporator's thermostatic bath; (**g**) rotary flask (which will contain the micronized sucrose at the beginning of the process and the proliposomes at the end of the process); and (**h**) receiving flask (which will contain condensed ethanol at the end of the process)

cap type, a hole should be made in the lid with a diameter equivalent to the diameter of the silicone hose.

3. Connect one end of a silicone hose to the vacuum pump (**E**) and the other end to the corresponding location on the rotary evaporator.
4. Verify that both the rotary flask (**F**) and the receiving flask (**G**) are properly clean and ready for use. Set aside.

Preparing the Ethanolic Bioactive Solution

A schematic representation of the production of the ethanolic bioactive solution is presented in Fig. 2.

1. With the assistance of a spatula with a 20 cm scoop, weigh 3.2 g of Phospholipon 90H (**A**) using an analytical balance and place

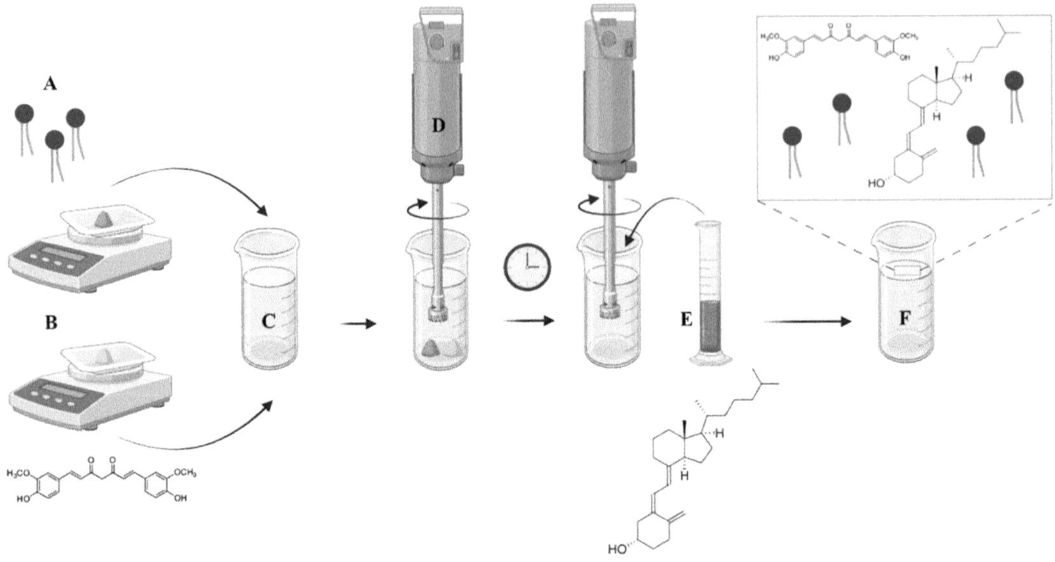

Fig. 2 Schematic representation of the production of the ethanolic bioactive solution containing curcumin and cholecalciferol through ultragitation

it in a crucible. Other phospholipids or phospholipids blends can be used depending on researcher's purpose.

2. With the assistance of a spatula with a 3 mm scoop, weigh 25 mg of curcumin (**B**) powder using an analytical balance and place it in a crucible.
3. Dissolve the curcumin and phospholipids in 100 mL of ethanol (**C**). It is suggested that the ethanol be contained in a 250 mL tall-form beaker.
4. With the aid of a disperser machine (**D**) operating at 13,000 rpm and 25 °C, homogenize the ingredients for 15 min. After the first minute of stirring, carefully pour 20 mL the cholecalciferol solution (**E**) from a graduated cylinder into the beaker containing the other ingredients. It is advised to pour the cholecalciferol solution gradually to ensure a homogeneous mixture with the other ingredients and to prevent premature degradation of the bioactive compound. At the end of the process, an ethanolic solution of bioactives is obtained (**F**).

Starting the Experiment

1. Transfer the ethanolic solution of bioactives from the 250 mL tall-form beaker into a 250 mL Schott flask and place the silicone hose inside, ensuring it passes through the screw cap and reaches the bottom. Seal the Schott flask tightly.

2. With the assistance of a spatula with a 20 cm scoop, weigh 2 g of micronized sucrose using an analytical balance and place it in a crucible. Transfer the content into the rotary flask.
3. Connect the appropriate receiving flask provided with the instrument to the end of the vertical condenser with a clamp. This flask will contain the condensed ethanol postevaporation.
4. Turn on the thermostatic bath and wait for it to reach 10 ± 2 °C. If necessary, place ice packs inside it.
5. Add deionized water to the rotary evaporator's thermostatic bath (F), turn it on and wait for it to reach 45 ± 5 °C.
6. Seal all possible vacuum leakage points with parafilm (mainly those where there are junctions containing silicone hoses).
7. Adjust the rotary evaporator so that the content of the volumetric flask is fully submerged in the hot water and turn on the rotation. Here, a moderate rotation is advisable. Rotations that are too fast or too slow may affect the contact of the ethanolic solution with the micronized sucrose or hinder the evaporation of ethanol.
8. Turn on the vacuum pump and check if the vacuum is being formed. If it is not being formed, check if all possible outlets are properly sealed with parafilm.
9. If the vacuum has been properly formed, turn on the peristaltic pump, set the flow to 4 mL/min, and verify if the ethanolic solution of bioactives is effectively being pumped into the rotary evaporator and dripped onto the micronized sucrose continuously.
10. Ensure that ethanol is being evaporated and condensed throughout the process. Condensation occurs when the evaporated ethanol (at 45 °C) comes into contact with the cold surface of the condenser (at 10 °C). The liquid ethanol is collected in the receiving flask connected to the condenser.
11. Conduct the injection process of the ethanolic solution of bioactives until it is completely depleted. Turn off the peristaltic pump and let the drying process continue for an additional 40–50 min to ensure complete removal of ethanol.
12. At the end of this time, stop the rotation of the thermostatic bath of the rotary evaporator, open the purge of the vacuum pump to release the vacuum, and turn it off.
13. Turn off the thermostatic bath of the rotary evaporator.
14. If no more repetitions of the process will be done, also turn off the thermostatic bath that is set to 10 °C.
15. Uncouple the receiving flask containing condensed ethanol and dispose of it properly. This ethanol likely contains traces of curcumin that are carried by the ethanol throughout the process.

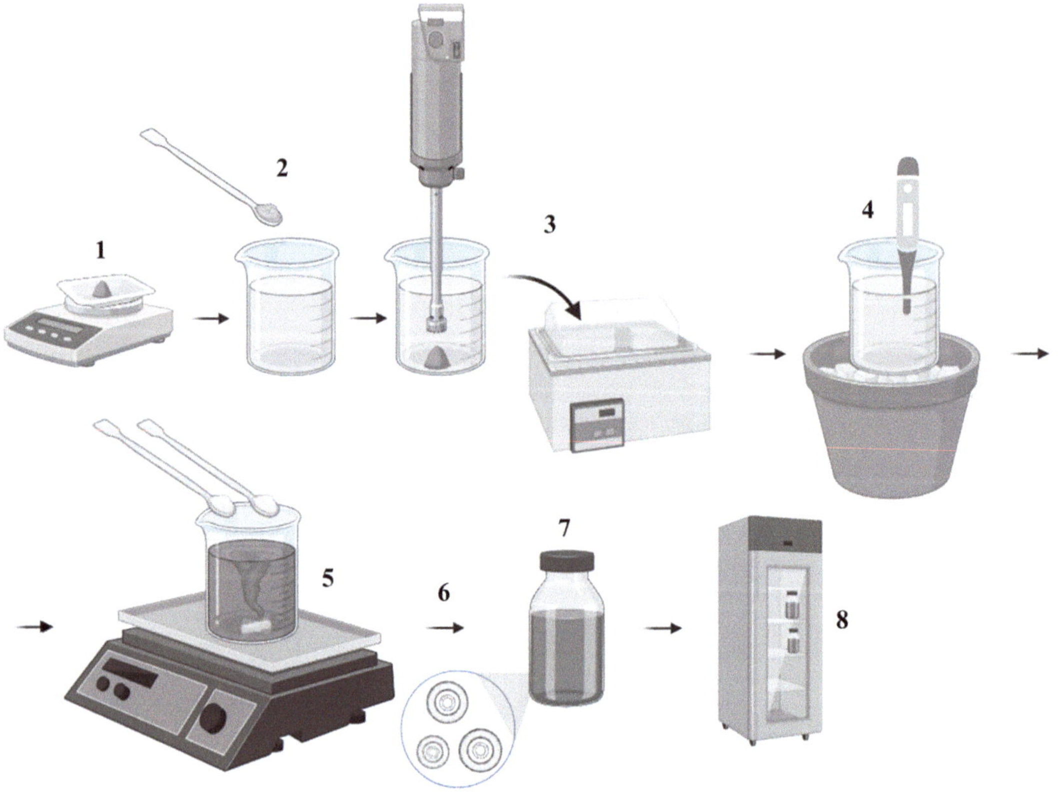

Fig. 3 Schematic representation of the production of liposome dispersions through the method of hydration of the proliposomes

16. Uncouple the rotary flask from the rotary evaporator and scrape its entire inner surface with a spatula to collect the proliposomes.
17. Transfer the proliposomes to amber flasks and store in a desiccator at 25 °C with controlled humidity.
18. Store the proliposomes for a maximum of 2 months.

3.2 Production of Curcumin/Cholecalciferol-co-Loaded Liposome Dispersions

A schematic representation of the production of liposome dispersions is presented in Fig. 3.

1. With the assistance of a spatula with a 3 mm scoop, weigh 0.01 g (10 mg) of xanthan gum and 0.09 g (90 mg) of guar gum separately into crucibles and set aside.
2. With the assistance of a spatula with a 20 cm scoop, weigh 2 g of proliposomes into a 250 mL beaker and pour in 100 mL of deionized water.
3. Place the beaker containing the ingredients in a hot water bath at 65 °C. While the beaker is in the hot water, homogenize the

system using an Ultra-Turrax at 15,000 rpm for 5 min. Liposome dispersions form from the contact of proliposomes with the buffer (deionized water) and the addition of energy (in this case, provided by mechanical agitation). The temperature of 65 °C was chosen because it is above the phase transition temperature of the chosen phospholipid for production (Phospholipon 90H). If another phospholipid is chosen, this temperature should be known.

4. Transfer the system to an ice bath to reduce the temperature to 25 °C. The ice bath can be made by adding ice to water in a container where the entire liposome dispersion can be surrounded by chilled water.
5. Place the solution on a magnetic stirrer at 3600 rpm at 25 °C and add the xanthan and guar gums. Ensure that the gums are added after the vortex formation begins to prevent lump formation. Let it stir for 20 min.
6. With the assistance of a spatula with a 3 mm scoop, weigh 20 mg (0.02 g) of sodium benzoate and transfer to the mixture. Let it stir for additional 10 min.
7. Transfer the liposome dispersions to amber flasks.
8. Stock the liposome dispersions at 10 °C for a maximum of 2 weeks.

3.3 Curcumin Content Analysis

Constructing a Curcumin Standard Curve by Spectrophotometry

1. Weigh an appropriate amount of curcumin and dissolve it in ethanol to make a stock solution of known concentration (e.g., 100 μg/mL).
2. Prepare a series of standard solutions by diluting the stock solution with ethanol to obtain concentrations of 1, 2, 3, 4, 5, and 6 μg/mL. Use the following dilutions:

   1 μg/mL: 100 μL (0.1 mL) of stock solution in 10 mL ethanol.

   2 μg/mL: 200 μL (0.2 mL) of stock solution in 10 mL ethanol.

   3 μg/mL: 300 μL (0.3 mL) of stock solution in 10 mL ethanol.

   4 μg/mL: 400 μL (0.4 mL) of stock solution in 10 mL ethanol.

   5 μg/mL: 500 μL (0.5 mL) of stock solution in 10 mL ethanol.

   6 μg/mL: 600 μL (0.6 mL) of stock solution in 10 mL ethanol.
3. Turn on the UV-Vis spectrophotometer and set it to the wavelength appropriate for curcumin (e.g., 425 nm).
4. Calibrate the spectrophotometer using ethanol as the blank.
5. Measure and record the absorbance of each standard solution at the selected wavelength.

6. Plot the absorbance values (y-axis) against the concentrations of the standard solutions (x-axis) using a graphing software (Microsoft Excel®, OriginPro®, for example).
7. Draw the best fit line through the data points.
8. Determine the equation of the line ($y = ax + b$) where "y" is the absorbance, "x" is the concentration, "a" is the slope, and "b" is the y-intercept. Ensure the standard curve has an R^2 value higher than 0.99 to accurate quantification.

Performing the Analysis

1. Ensure all materials and equipment are ready, including anhydrous ethanol, the liposome dispersions, and the spectrophotometer. It is advisable to take the liposome dispersion samples out of the refrigerator in advance so that the analysis can be performed with all samples at 25 °C.
2. Add 1900 μL (1.9 mL) of anhydrous ethanol to a 2 mL Eppendorf tube.
3. Add 100 μL of liposome dispersion to the tube with ethanol.
4. Vigorously shake the mixture to ensure thorough mixing. It will be possible to verify the lipid/gum phase being separated from the dispersion. Remove the precipitate using a filter membrane (0.45 μm, Ø = 13 mm).
5. Transfer the mixture to a suitable cuvette for spectrophotometry. For absorbance measurements in the visible region, glass or quartz cuvettes are acceptable. Do not use plastic cuvettes, as they are not suitable for use with organic solvents as ethanol.
6. Calibrate the spectrophotometer using ethanol as the blank and perform the analysis.
7. Using the recorded absorbance value, determine the curcumin concentration referring to the previously constructed standard curve of pure curcumin in ethanol. For this, replace the absorbance value obtained from the sample in "y"; the curcumin concentration will be the respective obtained in "x."
8. Clean and store all equipment and materials used in the procedure.
9. Dispose of any waste materials properly.

3.4 Cholecalciferol Content Analysis

Constructing a Cholecalciferol Standard Curve Using High-Performance Liquid Chromatography (HPLC)

1. Weigh an appropriate amount of cholecalciferol and dissolve it in methanol-acetonitrile 90:10 v/v to make a stock solution of known concentration (e.g., 1000 μg/mL).

2. Prepare a series of standard solutions by diluting the stock solution with the same solvent to obtain concentrations of 0.5, 1.0, 2.5, 5.0, 10.0, and 15.0 μg/mL. Use the following dilutions:

   0.5 μg/mL: 5 μL (0.005 mL) of stock solution in 10 mL solvent.

   μg/mL: 10 μL (0.01 mL) of stock solution in 10 mL solvent.

   2.5 μg/mL: 25 μL (0.025 mL) of stock solution in 10 mL solvent.

   5.0 μg/mL: 50 μL (0.05 mL) of stock solution in 10 mL solvent.

   10.0 μg/mL: 100 μL (0.1 mL) of stock solution in 10 mL solvent.

   15.0 μg/mL: 150 μL (0.15 mL) of stock solution in 10 mL solvent.

3. Turn on the HPLC system and allow it to equilibrate.
4. Set the appropriate HPLC parameters for cholecalciferol analysis (e.g., column type, mobile phase composition, flow rate, detection wavelength). In our previous studies, we used the following parameters, which can serve as a reference: a Shim-Pack VP-ODS column (4.6 μm, 0.46 × 25 cm) from Shimadzu, Japan; a mobile phase constituted of methanol-acetonitrile in a 90:10 v/v ratio; a flow rate of 1.6 mL/min; and a detection wavelength set at 265 nm. It is recommended that all parameters remain the same, except for the column, which can be selected at the researcher's discretion. However, the mobile phase must to be the same as the one used to construct the calibration curve.
5. Inject each standard solution into the HPLC system and record the area under the curve (AUC) for the cholecalciferol peak.
6. Plot the AUC values (y-axis) against the concentrations of the standard solutions (x-axis) using a graphing software (Microsoft Excel®, OriginPro®, for example).
7. Draw the best fit line through the data points.
8. Determine the equation of the line ($y = ax + b$) where y is the AUC, x is the concentration, a is the slope, and b is the y-intercept. Ensure the standard curve has an R^2 value higher than 0.99 to accurate quantification.

Performing the Analysis

Preparing the Samples

1. Ensure all materials and equipment are ready, including the HPLC apparatus, methanol, and acetonitrile.

2. Add 5 mL of methanol to a 15 mL Falcon tube.
3. Add 2 mL of liposome dispersion to the tube with methanol.
4. Vortex the mixture using a vortex mixer at 1250 rpm at 25 °C for 5 min to ensure thorough mixing.
5. Using a bath sonication, sonicate the vortexed solution at 25 °C for 10 min to further aid in the extraction process.
6. Using a refrigerate centrifuge, centrifuge the sonicated solution at 7400 × g at 4 °C for 5 min to separate the cholecalciferol from the lipid matrix.
7. Carefully collect the supernatant after centrifugation.
8. Transfer the supernatant to a 10 mL volumetric flask.
9. Bring the volume up to 10 mL with methanol.
10. Filter the methanolic extracts using nylon syringe filters (0.45 μm, Ø = 13 mm) to remove any particulate matter.
11. Transfer the filtered extracts into 2 mL HPLC vials. Store at 4 °C until use.

HPLC Analysis

1. Install the column in the HPLC system.
2. Set the column oven temperature to 35 °C.
3. Prepare the mobile phase using methanol and acetonitrile in a ratio of 9:1 (v/v).
4. Set the HPLC system to a constant flow rate of 1.6 mL/min.
5. Set the detection wavelength to 265 nm.
6. Inject the sample into the HPLC system using the automatic injector.
7. Run the HPLC for 10 min per sample.
8. Identify cholecalciferol by comparing the retention time with the standard. The retention time varies between cholecalciferol standards and different types of columns. The retention time must be the same as that obtained during the construction of the calibration curve.
9. Measure the area under the curve (AUC) for each sample related to the cholecalciferol peak.
10. Using the recorded AUC value, determine the cholecalciferol concentration referring to the previously constructed standard curve of pure cholecalciferol in methanol. For this, replace the AUC value obtained from the sample in "y"; the cholecalciferol concentration will be the respective obtained in "x."
11. Clean the HPLC system according to the manufacturer's instruction and store all materials used in the procedure.
12. Dispose of any waste materials properly.

References

1. Subramani T, Ganapathyswamy H (2020) An overview of liposomal nano-encapsulation techniques and its applications in food and nutraceutical. J Food Sci Technol 57(10): 3545–3555. https://doi.org/10.1007/s13197-020-04360-2
2. Emami S, Azadmard-Damirchi S, Peighambardoust SH, Valizadeh H, Hesari J (2016) Liposomes as carrier vehicles for functional compounds in food sector. J Exp Nanosci 11(9):737–759. https://doi.org/10.1080/17458080.2016.1148273
3. Shukla S, Haldorai Y, Hwang SK, Bajpai VK, Huh YS, Han YK (2017) Current demands for food-approved liposome nanoparticles in food and safety sector. In: Frontiers in microbiology, vol 8, Issue DEC. Frontiers Media S.A. https://doi.org/10.3389/fmicb.2017.02398
4. Fu DW, Xu H, Sun RQ, Liu XL, Ji Z, Zhou DY, Song L (2024) Engineering marine phospholipid nanoliposomes via glycerol-infused proliposomes: mechanisms, strategies, and versatile applications in scalable food-grade nanoliposome production. Food Chem 448: 139030. https://doi.org/10.1016/j.foodchem.2024.139030
5. Wang F, Pu C, Liu M, Li R, Sun Y, Tang W, Sun Q, Tian Q (2022) Fabrication and characterization of walnut peptides-loaded proliposomes with three lyoprotectants: environmental stabilities and antioxidant/antibacterial activities. Food Chem 366:130643. https://doi.org/10.1016/j.foodchem.2021.130643
6. Crowe LM, Crowe JH, Rudolph A, Womersley C, Appel L (1985) Preservation of freeze-dried liposomes by Trehalose. In: Archives of biochemistry and biophysics, vol. 242, issue 1
7. Elhissi AMA, Islam MA, Arafat B, Taylor M, Ahmed W (2010) Development and characterisation of freeze-dried liposomes containing two anti-asthma drugs. Micro Nano Lett 5(3):184–188. https://doi.org/10.1049/mnl.2010.0032
8. Elhissi AMA, Karnam KK, Danesh-Azari M-R, Gill HS, Taylor KMG (2006) Formulations generated from ethanol-based proliposomes for delivery via medical nebulizers. J Pharm Pharmacol 58(7):887–894. https://doi.org/10.1211/jpp.58.7.0002
9. Chaves MA, Pinho SC (2019) Curcumin-loaded proliposomes produced by the coating of micronized sucrose: influence of the type of phospholipid on the physicochemical characteristics of powders and on the liposomes obtained by hydration. Food Chem 291:7–15. https://doi.org/10.1016/j.foodchem.2019.04.013
10. Chaves MA, Pinho SC (2020) Unpurified soybean lecithins impact on the chemistry of proliposomes and liposome dispersions encapsulating vitamin D3. Food Biosci 37: 100700. https://doi.org/10.1016/j.fbio.2020.100700

Chapter 9

Production of Pickering Emulsions Stabilized by Protein Microgels

Jessica Thaís do Prado Silva, Elisa Franco Ribeiro, Tiago Carregari Polachini, and Vânia Regina Nicoletti

Abstract

Pickering emulsions (emulsions stabilized by particles) are receiving increasing interest for application in the food field due to their high physical stability and clean label appeal. Among the food-grade particles used for Pickering stabilization, protein microgels stand out due to their elastic properties and surface-active character. Protein solutions can be converted to protein microgel dispersions by controlled aggregation of the macromolecules (bottom-up) or disruption of bulk gel (top-down) techniques. The microgel dispersion is then used as the continuous phase to produce Pickering emulsions. Continuous and disperse phases are homogenized either by low-energy (membrane emulsification and microfluidics) or high-energy methods (rotor-stator homogenization, high-pressure homogenization, and sonication), forming an oil/-water emulsion. In addition, the encapsulation of lipophilic bioactive compounds may be accomplished by solubilizing these substances in the disperse phase. Here, we describe the main methodologies to produce protein microgel particles and how to use such particles to stabilize Pickering emulsions, respectively.

Key words Thermal denaturation, Protein crosslinking, Soft particles, Emulsification methods, Stable emulsion

1 Introduction

In Pickering emulsions, nano- or microparticles are placed at the oil/water interface, creating a physical barrier that prevents droplets from merging upon collision. These systems are well known for their high physical stability. However, more recently, Pickering emulsions are being explored as a clean label strategy for producing food emulsions, since they do not require the use of low molecular weight surfactants [1]. Despite being at the early stages of development in the food field, Pickering emulsions have also proven to be efficient carrier systems for the encapsulation of lipophilic bioactive compounds [2].

Samantha Cristina de Pinho and Carmen S.F. Trindade (eds.), *Bioactives Encapsulation: Food Applications*, Methods and Protocols in Food Science, https://doi.org/10.1007/978-1-0716-4538-3_9,

Microgel particles are kinetically stable entities held together by covalent bonding, with the ability to trap solvents within their crosslinked polymer network and occupy an intermediate position between colloidal particles and polymer molecules [3–5]. When in aqueous media, microgel particles present a roughly spherical shape. They form a core-shell morphology with a denser core and a loosely cross-linked shell, in which dangling polymers are located at the outermost part. Microgels made from crosslinked polymers are the most recognized soft particles for Pickering stabilization.

The role of soft particles as emulsion stabilizers is somehow different from that of hard classical particles. For well-defined hard particles, the adsorption at the interface is mainly ruled by their contact angle, forming a strong physical barrier against coalescence (classical Pickering emulsion). In contrast, microgel adsorption is ruled by the surface-active character of their dangling chains, whereas their stabilizing ability is linked to the viscoelasticity of the formed interfacial layer [5, 6].

After the emulsification process, microgels adsorbed at the interface stretch as much as possible, sometimes reaching twice their diameter when in bulk. This deformation is driven by the decrease in free energy (interfacial tension), and it is counterbalanced by their intrinsic viscoelasticity, which is directly related to their crosslinking degree. In this scenario, microgels assume a unique structure with a protruding core and a wide shell made of long ramifications that cover the remaining part of the interface, forming a "fried-egg-like" structure [7, 8].

Regular emulsification processes used to prepare conventional emulsions can also be applied to preparing Pickering emulsions [9, 10]. Nonetheless, it should be considered that emulsification energy can affect the deformation of microgels at the oil/water interface. According to Destribats et al. [6], high emulsification energy allows microgels to assume several possible configurations, enhancing the adsorption of their dangling chains and resulting in strong microgel deformation at the interfacial plane. Moreover, a larger interfacial area is created by applying a high-shear mixing process, and in such conditions, microgels have enough space to deform laterally, although their density at the interface is low, which may promote the occurrence of bridging flocculation (Fig. 1a). On the other hand, when using low-shear emulsification, the deformation dynamics of microgel particles are too slow compared to the droplet's recombination rate, and the interfacial area created during homogenization is small. In this way, microgels neither have time nor space to extend. Thus, a highly dense interfacial layer is created, hindering bridging between droplets, and consequently avoiding flocculation (Fig. 1b).

Protein microgels combine the high desorption energy of Pickering particles with the ability of protein molecules to adsorb at the oil/water interface. These characteristics combined with the high

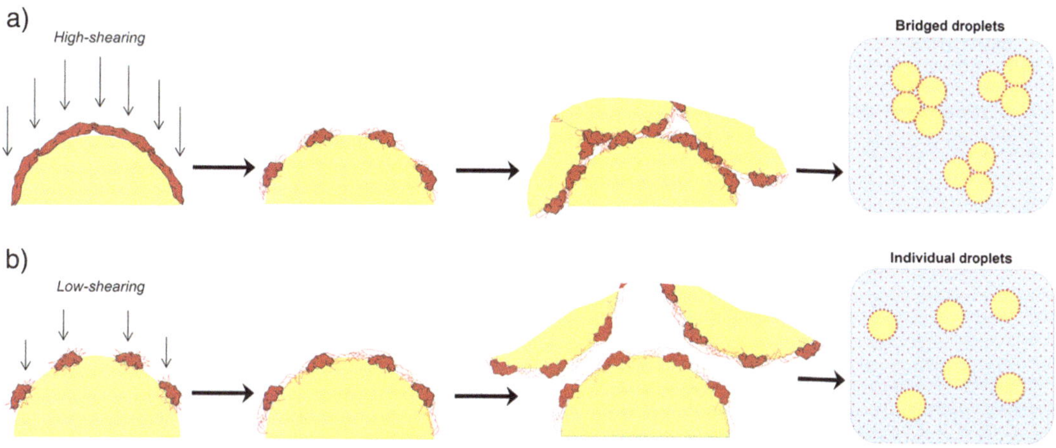

Fig. 1 Effects of high-shear (**a**) and low-shear (**b**) mixing on microgel configuration

nutritional value of commonly available proteins make them an advantageous stabilizer for producing food-grade Pickering emulsions. Furthermore, since most of these proteins are more compatible with water than with oil, their derived microgels are better suited (or designed) to stabilize oil-in-water emulsions [11, 12].

This chapter will present the most used methodologies for preparing food-grade Pickering emulsions stabilized by protein microgels. This includes two steps: the production of microgels and the emulsification process involving the microgel dispersion and an oil phase.

2 Materials

2.1 Production of Protein Microgels

Protein isolate or concentrate from an animal or vegetal source. Deionized water or phosphate buffer is used for preparing the protein solution at a desired pH or ionic strength. An antimicrobial agent such as sodium sorbate may be added to the protein solution to prevent microbial growth during microgels' storage over time. Optionally, crosslinking agents can be applied to produce bulk gels. The most used crosslinking agents for protein microgels are organic acids, such as tannic acid, citric acid, and genipin [13, 14]; or enzymes such as transglutaminase [15, 16].

2.2 Production of Pickering Emulsions Stabilized by Protein Microgels

The protein microgel dispersion is used as the continuous phase, and a vegetal oil (soy, sunflower, rapeseed, etc.) containing the bioactive compound to be encapsulated is applied as the dispersed phase.

3 Methods

3.1 Preparation of the Protein Solution

Protein isolate or concentrate is mixed with deionized water or phosphate buffer under magnetic stirring for 2 h. Afterward, the protein solution is stored at low temperatures (~4 °C) for 12 h, to guarantee complete hydration of the molecules.

Parameters such as protein concentration, pH, and ionic strength of the solution depend on the denaturation behavior of the protein used, which in turn depends on its molecular structure.

3.2 Production of Protein Microgels

Protein microgel particles can be produced by two different approaches: "bottom-up" or "top-down" techniques.

3.2.1 Bottom-Up Technique

In the bottom-up technique, the gelation of proteins takes place at low protein concentration and/or significant electrostatic repulsion between molecules. This enables a slow and controlled aggregation of protein molecules to form fractal gel particles [12].

A simple heating protocol is required for producing microgels by this technique. The protein solution, prepared as described in Subheading 3.1, is transferred to a thermal water bath, without stirring, during a determined period. At temperatures above 80 °C and at a certain pH range, protein molecules aggregate, promoting the formation of relatively spherical microgel particles composed of densely packed chains. The microgel suspension is then cooled either by running water or in an ice bath to stop the denaturation process.

This approach has shown excellent results, especially when applied to milk proteins, such as whey protein isolate (WPI) and β-lactoglobulin. Usually, WPI microgels are prepared with a 4 wt.% WPI solution at a pH 5.8, which is heated at 85 °C for 15 to 45 min. Throughout this procedure, it is possible to obtain particles with mean diameters varying from 140 nm to 348 nm [17–19]. For β-lactoglobulin microgels, 4 wt.% β-lactoglobulin solution at a pH range of 5.8–6.9 is heated at 85 °C from 20 min to 15 h, giving rise to particles with mean diameters varying from 140 nm to 266 nm [20, 21].

Although the bottom-up technique for producing protein microgels has been dominated by milk proteins, good results have also been reported for proteins from plant sources. For instance, soy protein isolate microgels were obtained by heating a 2 wt.% soy protein solution at 95 °C for 15 min, resulting in particles of 92 nm mean diameter [22].

3.2.2 Top-Down Technique

In the top-down technique, a bulk gel is formed by a crosslinking approach, followed by disruption into microgel particles via mechanical forces. In other words, the controlled fragmentation of a bulk gel is performed [12]. Here, the most used crosslinking methods for producing bulk gels are thermal denaturation, chemical crosslinking, and enzymatic crosslinking. The mechanical force applied to disrupt the bulk gel is usually provided by a high-pressure homogenizer or an ultrasonication probe. The top-down approach is applied to a broad class of proteins, whether they are from animal or plant sources.

For thermal denaturation, a thermal treatment is applied to the protein solution. The procedure is quite similar to that performed in the bottom-up technique; however, in this case, the starting protein solution is more concentrated. For instance, soy protein microgels are prepared by heating a 10 wt.% soy protein solution (pH 7.0) for 30 min at 85 °C. Afterward, the bulk gel obtained is processed in Ultra-Turrax followed by high-pressure homogenization at 30 bar in the first stage and 300 bar in the second stage, giving rise to soy protein microgels with a 395-nm mean diameter [23]. WPI microgels are prepared by heating a 10 wt.% WPI solution (pH 7.0) at 85 °C for 30 min. Next, the bulking gel is processed in an ultrasonication probe (20 KHz) at 40% amplitude for 5 to 40 min, forming microgel particles ranging from 43 nm to 34 nm [24]. For pea protein microgel production, a 12.75 wt.% pea protein solution (pH 7.0) is heated at 90 °C for 1 h. The bulk gel is disrupted in a blender and high-pressure homogenizer at 250 and 50 bar in the first and second stages, respectively, producing microgel particles of 232 nm mean diameter [25]. Li et al. [26] applied a similar procedure for the production of egg white protein (EWP) microgels. A 6.25 wt.% protein solution (pH 7.0) was heated at 90 °C for 30 min, followed by disruption by a high-pressure homogenizer at 300 bar. They obtained EWP microgels of around 100 nm.

In chemical and enzymatic crosslinking, the bulk gel is formed by linking amino acid chains of proteins by noncovalent or covalent bonds, forming tridimensional networks [27]. For food-grade protein microgels, these linkages are facilitated by organic acids (chemical crosslinking) or transglutaminase (enzymatic crosslinking). The process consists of introducing the crosslinking agent to the protein solution after the hydration time. In some cases, a heat treatment is also applied to promote a partial unfolding of protein molecules, in a way to optimize the crosslinking reaction. Silva et al. [13] used citric acid and tannic acid to produce chemically crosslinked WPI bulk gels. A 4 wt.% WPI solution was mixed with the crosslinker at a 3:1 (crosslinker:protein) molar ratio and submitted to a thermal treatment (80 °C/15 min). The obtained bulk gels were treated in an ultrasonication probe (20 Hz) for 3 min, forming particles of 185 nm and 257 nm mean diameter for tannic acid- and citric acid-crosslinked WPI microgels, respectively. Guo et al. [15] used transglutaminase to produce enzymatically crosslinked soy protein bulk gel. For this purpose, a 10 wt.% soy protein solution was first heated (80 °C/20 min) to promote the opening of protein molecules. Transglutaminase (20 U/g soy protein) was then added to the heated solution, which was kept at 40 °C for 4 h for the enzymatic reaction. Finally, the bulk gel passed through a high-pressure homogenizer at 500 bar, giving rise to microgel particles around 141- to 158-nm mean diameter. A similar procedure was applied to produce transglutaminase-crosslinked sea bass protein (SBP) bulk gel by Zhang et al. [28]. A 10 wt.% SBP solution was heat treated (80 °C/20 min), followed by transglutaminase treatment (20 U/g SBP) for 4 h at 45 °C. The bulk gel was then homogenized in an Ultra-Turrax and disrupted in a high-pressure homogenizer at 10000 psi for three cycles. The microgels obtained presented a 300- to 500-nm mean diameter.

3.3 Production of the Emulsions

The homogenization process of two immiscible phases resulting in small dispersed droplets in a continuous phase is called emulsification. In general, this consists of the reduction of oil droplet size, providing a significant improvement in the emulsion's physical stability.

The same processes used to prepare conventional emulsions are also applied to prepare Pickering emulsions. The most common homogenization techniques to formulate Pickering emulsions stabilized by protein microgels are high-shear mixing, such as when using rotor-stator homogenization, high-pressure homogenization, or sonication. Membrane emulsification and microfluidic emulsification are novel low-shear mixing operations that have been explored for better control over the droplet microstructure and, consequently, the quality of the final emulsions.

3.3.1 High-Shear Mixing Methods

- *Rotor-Stator Homogenization*

The rotor-stator homogenizer features a rotating metal shaft (the rotor) inside a stationary metal casing (the stator) (Fig. 2). The homogenization method consists of circulating the sample through the stator as the rotor rotates. The rapid fluid movement caused by the rotor-stator geometry creates a suction effect that draws the sample into the small space between the rotor and stator. Due to an extreme change in the fluid velocity in the small flow area, samples are subjected to very high shear forces, providing a reduction in droplet sizes of the emulsion. In addition to the rotation speed, the emulsion droplet size is also influenced by the homogenization time. The droplet size distribution in Pickering emulsions is broad (from a few micra to hundreds of micra) since the rotation speeds generally used are in a range of 5000 to 30,000 rpm, whereas the homogenization times range from 30 seconds to a few minutes [9]. Liu et al. [29] evaluated the potential of egg white microgels as Pickering emulsion stabilizers using a bottom-up methodology to prepare the microgels, as well as a high-speed homogenizer (10,000 rpm for 1 min) to produce the emulsions. The authors obtained high apparent viscosity emulsions with elastic gel properties and small particle size, which were responsible for preventing droplet aggregation and flocculation.

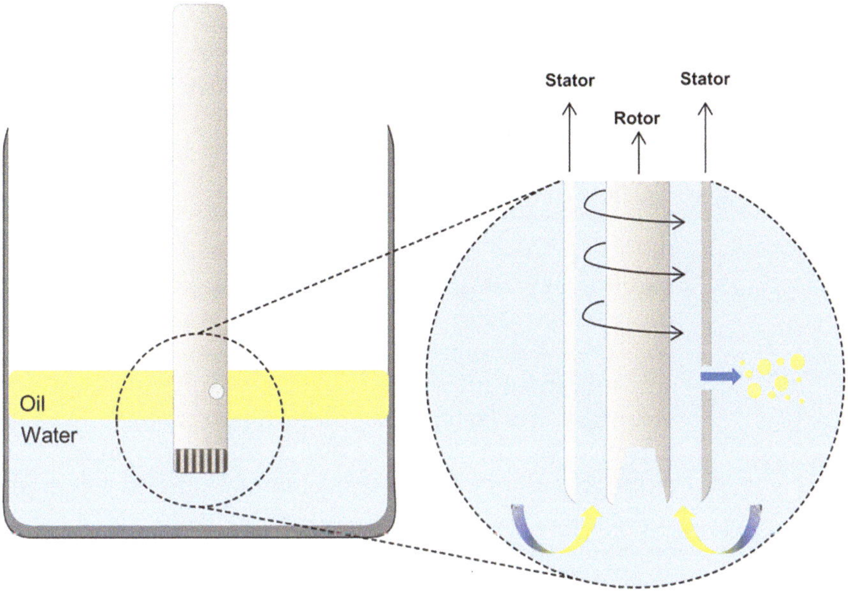

Fig. 2 Schematic representation of a rotor-stator homogenizer

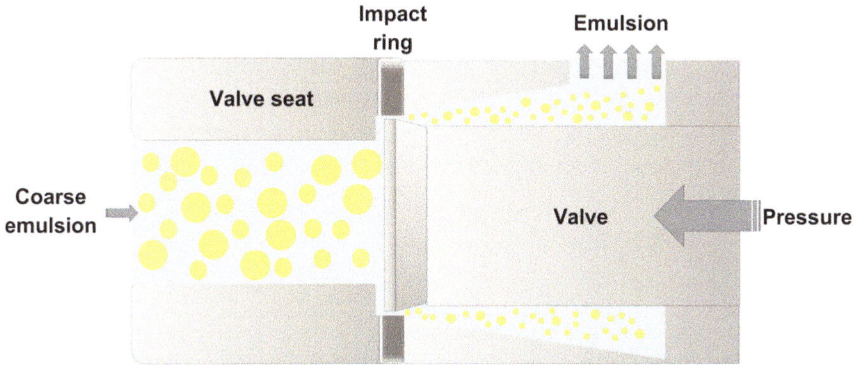

Fig. 3 Schematic representation of a high-pressure homogenizer

- *High-Pressure Homogenization*

The high-pressure homogenizer consists of a high-pressure pump and a homogenizing nozzle (Fig. 3). Based on the same principle as conventional homogenization, the difference with this technology is the pressure of up to 400 MPa achieved for homogenization. This technique allows a fluid to pass through a narrow gap in the homogenizer valve, deforming the suspended materials in the fluid through large mechanical forces and elongation stresses, which induce emulsification. To obtain a fine emulsion at the homogenizer outlet, a previous pre-emulsification step is suggested, which can be carried out using a rotor-stator or vortex mixer. Homogenization cycles can be repeated as many times as necessary to reduce droplet size to a desired value, as Pickering emulsion droplets range from hundreds of nanometers to hundreds of micrometers. Silva et al. [30] produced previous coarse emulsions with whey protein microgels using a high-speed blender at 11,000 rpm for 1 min. Then, a colloid mill (rotor-stator homogenizer) or a microfluidizer (high-pressure homogenizer) was used to obtain the final emulsion by low shear or high shear, respectively. In this step, the homogenization conditions were 2 min at 15,000 rpm using a rotor-stator homogenizer and 400 bar for five passes using a high-pressure homogenizer. Higher viscosity emulsions were obtained by high-pressure homogenization, probably due to an increased flocculation in the emulsions when this technique was applied. Authors reported that the high energy provided by the high-pressure homogenizer made the flexible microgels adopt a deformed configuration at the interfacial plane, and subsequently, a lower density of microgels was necessary to cover the interface. However, this implies that, after a relaxation time, poorly covered droplets make up the system, which promotes the occurrence of flocculation phenomena.

- *Ultrasonic Emulsification*

The emulsification process using ultrasound comprises two stages: in the first stage, the large droplets of the dispersed phase are formed in the continuous phase. In the second stage, the ultrasound device generates acoustic cavitation, leading to the collapse of air bubbles in the system (Fig. 4). Consequently, physical

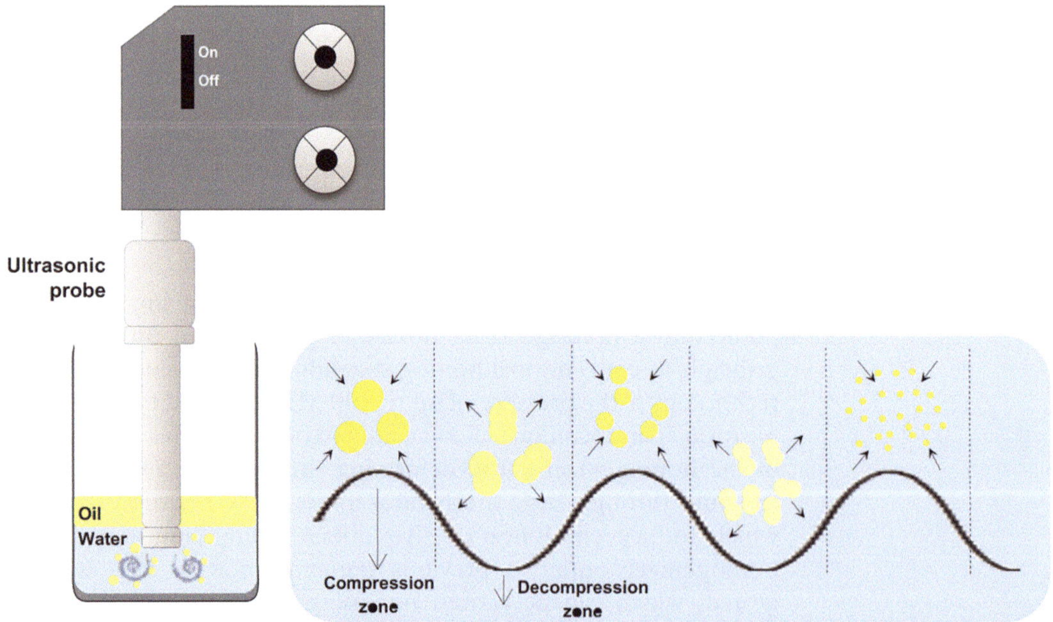

Fig. 4 Schematic representation of ultrasound emulsification

effects such as microjets, high temperature, high-pressure shock waves up to 1 GPa, and high local shear forces can gradually break down the large oil droplets into smaller ones, facilitating the formation of stable emulsions [31, 32]. The main parameters of the ultrasonic process that influence the droplet size are ultrasound frequency, power, and emulsification time. Generally, for obtaining the smallest droplet sizes, a low frequency range is used (from 20 to 40 kHz), with power ranges from tens to hundreds of watts. As the sonication power increases, the shear forces produced on the collapse of bubbles increase, inducing the disruption of oil droplets. Furthermore, a few minutes are enough for emulsification and a large range of volumes can be used in the system [9, 31, 33]. Ultrasound has been frequently used as an emulsification system for producing emulsions stabilized by protein microgels [13, 23]. Benetti et al. [23] applied sonication for 3 min to decrease the droplet size of a coarse emulsion previously obtained by mechanical stirring at 9000 rpm. In this work, the authors compared sonication with high-pressure homogenization for preparing emulsions with soybean protein isolate microgels. More intense shear-thinning behavior was presented by the emulsions when ultrasonication was applied. This is probably due to the differences in the arrangement of microgels at the interface in each emulsification technique used.

3.3.2 Low-Shear Mixing Methods

- *Membrane Emulsification*

Pickering emulsions can be produced by membrane emulsification in two different techniques: direct membrane emulsification (DME) (Fig. 5a) and premix membrane emulsification (PME) (Fig. 5b). Both methods consist of pushing the dispersed phase into the continuous one through the pores of a membrane that reduces the size of the droplets. Unlike the first method, PME injects an already pre-emulsified mixture as the dispersed phase. The stabilization of droplets against aggregation usually occurs with the use of an emulsifier—the microgels in the case of microgel-stabilized Pickering emulsions—previously dispersed in the continuous phase. Therefore, the type of membrane as well as the type and concentration of the emulsifier used are two parameters that could influence the properties of the obtained emulsions [31].

When using DME, some approaches are applied to detach the formed droplets in the continuous phase using low-shear forces on the surface of the membrane. When an additional mechanical agitation is required in the receiver chamber, it originates the stirred-cell membrane emulsification (SCME) (Fig. 5e). The use of a rotating or vibrating membrane generates rotational membrane emulsification (RME) (Fig. 5d) and vibrational membrane emulsification (VME), respectively. When a membrane tube in a flowing

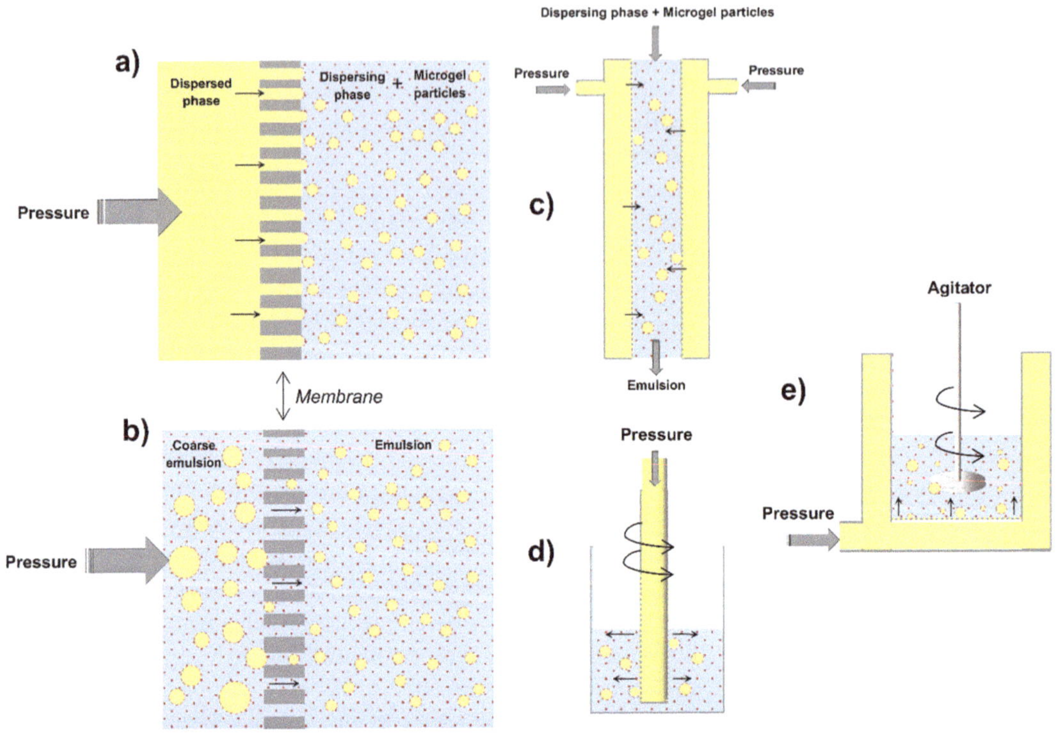

Fig. 5 Schematic representation of the main types of membrane emulsification techniques: (**a**) direct membrane emulsification (DME), (**b**) premix membrane emulsification (PME), (**c**) cross-flow membrane emulsification (XME), (**d**) rotational membrane emulsification (RME), and (**e**) stirred-cell membrane emulsification (SCME). (Adapted from Albert et al. [9])

continuous phase is used for passing the dispersed phase, it produces cross-flow membrane emulsification (XME) (Fig. 5c). Membranes can have uniform pore sizes that range from the nanometer to micrometer scale, with a tunable hydrophobicity. In addition to the membrane pore size, the emulsion droplet size is also controlled by the injection rate and agitation speed. Smaller and more uniform droplets are obtained with slower injection. On the other hand, the droplet size decreases as the agitation speed increases since it becomes easier to detach droplets from the membrane. In the case of low-speed agitation, larger droplet sizes are formed even before detaching from the membrane due to the low shear in the system [9].

- *Microfluidic Emulsification*

Aiming at the production of controllable monodisperse droplets, the microfluidic system has been used to prepare Pickering emulsions by drop-by-drop technology. The formation of droplets of one liquid into another is allowed by the laminar fluid flow through the microchannels of microfluidic devices with a particular geometry (Fig. 6). As in membrane emulsification, the shearing of

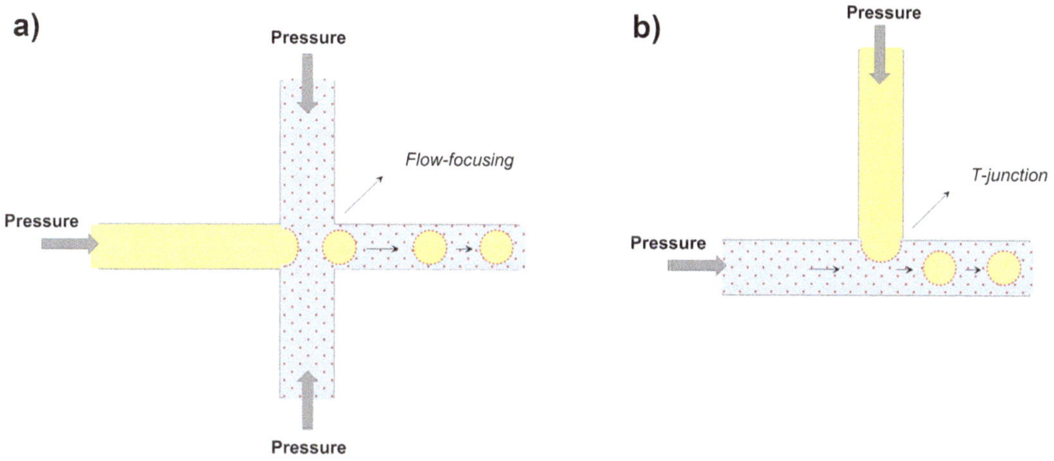

Fig. 6 Schematic representation of flow-focusing (**a**) and T-junction (**b**) microfluidic devices

the continuous phase leads to the formation of droplets. Droplet shape regulation occurs due to the ability to control the flow behavior during the stabilization of the dispersed droplets into the continuous phase. Different microfluidic devices are used for controlling the flow behavior, such as T-junction devices for cross-flow direction and flow-focusing devices for co-axial direction [9]. In the flow-focusing devices, the dispersed phase flows against two perpendicular streams of continuous phase that flow from both upper and lower channels (Fig. 6a). For the T-junction system, the dispersed phase flows through a small orifice into the continuous phase following a perpendicular channel (Fig. 6b). In this system, both phases flow through the orifice, inducing the formation of droplets [34]. The microfluidic technique allows controlling the droplet sizes as well as the frequency of droplet generation by tuning the fluid flow rates and the size of the microchannels [35]. Silva et al. [17] investigated the stabilization of emulsions by tannic acid-crosslinked whey protein microgels using a 100 μm channel width T-junction and a 100 μm width meandering channel. The depth of all channels was 45 μm. Important insights into droplet–droplet interactions were reported in this study. Emulsion droplets experienced transient bridging flocculation when stabilized by soft microgels, and it led to an increase in stability against coalescence. The authors obtained control over the release frequency of the droplets into the collision channel at a constant frequency of $f_D = 349 \pm 25\ s^{-1}$ and droplet size distribution with a diameter of 70 ± 3 μm and low polydispersity. In addition, the adsorption time proved to be an important parameter for droplet stabilization. A greater amount of microgels was adsorbed as the adsorption time increased, probably due to the surface activity of protein microgel particles, which may lead to increased adsorption throughout the channel.

4 Summary

The production of Pickering emulsions stabilized by protein microgels is still in the initial stages of development. Even though such systems have demonstrated a great potential for applications in foods, whether serving as food structuring ingredients or as delivery systems for bioactive compounds. Protein microgels can be produced from a simple heat treatment protocol to a more complex methodology using crosslinking agents. Plant proteins can also be used for preparing microgels, which makes room for designing plant-based emulsions and foods. The choice of parameters such as pH and concentration of the protein solution, as well as temperature and time of the heat treatment, is related to the denaturation behavior of the selected protein. On the other hand, the homogenization technique used for producing Pickering emulsions stabilized by protein microgels, and therefore the corresponding process parameters, depends on the desired droplet size, which in turn is related to physical stability and, sometimes, to the rheological behavior of the emulsion. Pickering emulsions stabilized by protein microgels present countless possibilities, being a rich field to be exploited.

References

1. Berton-Carabin C, Schroën K (2019) Towards new food emulsions: designing the interface and beyond. Curr Opin Food Sci 27:74–81
2. Linke C, Drusch S (2018) Pickering emulsions in foods – opportunities and limitations. Crit Rev Food Sci Nutr 58:1971–1985
3. Destribats M, Lapeyre V, Wolfs M et al (2011) Soft microgels as Pickering emulsion stabilisers: role of particle deformability. Soft Matter 7:7689–7698
4. Tang C-H (2020) Globular proteins as soft particles for stabilizing emulsions: concepts and strategies. Food Hydrocoll 103:105664
5. Dickinson E (2017) Biopolymer-based particles as stabilizing agents for emulsions and foams. Food Hydrocoll 68:219–231
6. Destribats M, Wolfs M, Pinaud F et al (2013) Pickering emulsions stabilized by soft microgels: influence of the emulsification process on particle interfacial organization and emulsion properties. Langmuir 29:12367–12374
7. Deshmukh OS, Ende D van den, Stuart MC, et al (2015) Hard and soft colloids at fluid interfaces: adsorption, interactions, assembly & rheology. Adv Colloid Interf Sci 222:215–227
8. Fernandez-Rodriguez MA, Martín-Molina A, Maldonado-Valderrama J (2021) Microgels at interfaces, from mickering emulsions to flat interfaces and back. Adv Colloid Interf Sci 288:102350
9. Albert C, Beladjine M, Tsapis N et al (2019) Pickering emulsions: preparation processes, key parameters governing their properties and potential for pharmaceutical applications. J Control Release 309:302–332
10. Wu J, Ma GH (2016) Recent studies of pickering emulsions: particles make the difference. Small 12:4633–4648
11. Calabrese V, Courtenay JC, Edler KJ, Scott JL (2018) Pickering emulsions stabilized by naturally derived or biodegradable particles. Curr Opin Green Sustain Chem 12:83–90
12. Murray BS (2019) Microgels at fluid-fluid interfaces for food and drinks. Adv Colloid Interf Sci 271:1–7
13. Silva JT d P, JVM B, Alexandrino TT d B et al (2021) Whey protein isolate microgel properties tuned by crosslinking with organic acids to achieve stabilization of pickering emulsions. Food Secur 10:1296
14. Wei Z, Cheng Y, Zhu J, Huang Q (2019) Genipin-crosslinked ovotransferrin particle-

stabilized pickering emulsions as delivery vehicles for hesperidin. Food Hydrocoll 94:561–573
15. Guo J, Zhou Q, Liu YC et al (2016) Preparation of soy protein-based microgel particles using a hydrogel homogenizing strategy and their interfacial properties. Food Hydrocoll 58:324–334
16. Zhao Y, Wang P, Xu Y et al (2022) A study on the mechanisms of nanoparticle-stabilized high internal phase emulsions constructed by cross-linking egg white protein isolate with different transglutaminase concentrations. Food Secur 11:1765
17. Silva JT d P, Nicoletti VR, Schroën K, de Ruiter J (2022) Enhanced coalescence stability of droplets through multi-faceted microgel adsorption behaviour. J Food Eng 317:110850
18. Karbasi M, Askari G, Madadlou A (2021) Effects of acetyl grafting on the structural and functional properties of whey protein microgels. Food Hydrocoll 112:106443
19. Zamani S, Malchione N, Selig MJ, Abbaspourrad A (2018) Formation of shelf stable pickering high internal phase emulsions (HIPE) through the inclusion of whey protein microgels. Food Funct 9:982–990
20. Phan-Xuan T, Durand D, Nicolai T et al (2014) Heat induced formation of beta-lactoglobulin microgels driven by addition of calcium ions. Food Hydrocoll 34:227–235
21. Murphy RW, Zhu L, Narsimhan G, Jones OG (2018) Impacts of size and deformability of β-lactoglobulin microgels on the colloidal stability and volatile flavor release of microgel-stabilized emulsions. Gels 4:1–14
22. Zhu X-F, Zheng J, Liu F et al (2017) Freeze-thaw stability of pickering emulsions stabilized by soy protein nanoparticles. Influence of ionic strength before or after emulsification. Food Hydrocoll 74:37–45
23. Benetti JVM, Silva JT d P, Nicoletti VR (2019) SPI microgels applied to Pickering stabilization of O/W emulsions by ultrasound and high-pressure homogenization: rheology and spray drying. Food Res Int 122:383–391
24. Shen X, Zhao C, Guo M (2017) Effects of high intensity ultrasound on acid-induced gelation properties of whey protein gel. Ultrason Sonochem 39:810–815
25. Zhang S, Holmes M, Ettelaie R, Sarkar A (2020) Pea protein microgel particles as pickering stabilisers of oil-in-water emulsions: responsiveness to pH and ionic strength. Food Hydrocoll 102:105583
26. Li X, Yang Y, Murray BS, Sarkar A (2020) Combination of egg white protein and microgels to stabilize foams: impact of processing treatments. J Food Eng 275:1–11
27. Azeredo HMC, Waldron KW (2016) Cross-linking in polysaccharide and protein films and coatings for food contact – a review. Trends Food Sci Technol 52:109–122
28. Zhang L, Zaky AA, Zhou C et al (2022) High internal phase Pickering emulsion stabilized by sea bass protein microgel particles: food 3D printing application. Food Hydrocoll 131:107744
29. Liu B, Wang J, Qin S et al (2024) Microgels from egg white as novel stabilizers of pickering emulsion. J Food Eng 372:112009
30. Silva JT d P, Janssen A, Nicoletti VR et al (2023) Synergistic effect of whey proteins and their derived microgels in the stabilization of O/W emulsions. Food Hydrocoll 135:108229
31. Taha A, Ahmed E, Ismaiel A et al (2020) Ultrasonic emulsification: an overview on the preparation of different emulsifiers-stabilized emulsions. Trends Food Sci Technol 105:363–377
32. Wu WH, Eskin DG, Priyadarshi A et al (2021) New insights into the mechanisms of ultrasonic emulsification in the oil–water system and the role of gas bubbles. Ultrason Sonochem 73:105501
33. Ashokkumar M, Cavalieri F, Chemat F et al (2016) Handbook of ultrasonics and sonochemistry, vol 1. Springer, Singapore, p 1487
34. Liu Y, Li Y, Hensel A et al (2020) A review on emulsification via microfluidic processes. Front Chem Sci Eng 14:350–364
35. Wang W, Li BY, Zhang MJ et al (2023) Microfluidic emulsification techniques for controllable emulsion production and functional microparticle synthesis. Chem Eng J 452:139277

Chapter 10

Production of Pickering Emulsions Stabilized by Starch Particles

Giselle Vallim Corrêa Ramos, Santiago Ramírez-López, Cynthia Ditchfield, and Izabel Cristina Freitas Moraes

Abstract

Pickering emulsions have attracted recent research interest due to their superior stability and unique properties in delivering bioactives (such as polyphenols, anthocyanins, carotenoids, fatty acids, lipophilic and hydrophilic vitamins, among others) in food emulsions. Starch can be used as the material of stabilizing particles in this type of emulsion. However, native starch presents limitations due to its natural hydrophilicity and other characteristics that require simple or combined modifications to develop granules or particles that are suitable for stabilizing Pickering emulsions and their application in encapsulating bioactive compounds. This chapter provides two detailed and comprehensive methods, covering two starch modification methods: with octenyl succinic anhydride (OSA) and a combination of heat-moisture treatment (HMT) and antisolvent precipitation, the production of the Pickering emulsions, and its use in curcumin encapsulation.

Key words Cassava starch, Canihua starch, Curcumin, Heat-moisture treatment (HMT), Octenyl succinic anhydride (OSA), Antisolvent precipitation

1 Introduction

A Pickering emulsion is a type of emulsion that uses solid or colloidal particles as emulsifiers to anchor the oil–water interface through several mechanisms such as steric hindrance [1]. Encapsulating bioactive compounds in this type of emulsion involves some challenges, such as achieving the desired droplet size, potential for particle aggregation, and limitations in incorporating highly hydrophilic bioactive compounds, depending on the type of particle used [2]. However, their unique properties, like high stability and potential for controlled release, combined with the use of natural particles that are multifunctional, biocompatible, nontoxic, and

Samantha Cristina de Pinho and Carmen S.F. Trindade (eds.), *Bioactives Encapsulation: Food Applications*, Methods and Protocols in Food Science, https://doi.org/10.1007/978-1-0716-4538-3_10,

environmentally friendly, make this type of emulsion ideal for applications in food matrices, including encapsulation of bioactive compounds [3].

The natural particles can be obtained from renewable sources such as gelatin, cellulose, globular proteins, and starch [4]. Starch as a natural food ingredient has been successfully used in the preparation of Pickering emulsions after hydrophobic modification [5]. This interest derives from the fact that starch is considered GRAS (generally recognized as safe), nonallergenic, abundant, and low cost [6].

The properties of starch vary according to the ratio of amylose to amylopectin present in the granule, as well as its physical organization within the structure or degree of crystallinity and the size of the granule. Different botanical sources exhibit different physicochemical and functional properties. For example, canihua starch is a starch from a non-conventional source, with small size (0.8–1.0 μm) granules, approximately 31% of crystallinity, and 11–14% amylose. Canihua is widely cultivated in the Andes region [7]. In contrast, cassava starch has much larger granule sizes (2–32 μm), a crystallinity of 15–49%, and 0–30% of amylose [8]. Cassava is extensively produced in Brazil, and cassava starch is commercially available even in its extracted form.

An alternative to improve the physicochemical properties of native starches, as well as their emulsification capacity, is to modify them through chemical [9], physical [10], or enzymatic [11] treatments either separately or in combination. Modification with (octenyl succinic anhydride (OSA) is a chemical method, whereas the use of heat-moisture treatment (HMT) and antisolvent precipitation are considered physical methods for starch modification.

The modification of starch by OSA promotes amphiphilic properties and enhanced capability of lowering surface and interfacial tension; hence, it can be used as an emulsion stabilizer and encapsulating agent [12]. On the other hand, the combined HMT and nanoprecipitation methods produce particles with small sizes (~70–700 nm), rough surfaces, increased hydrophobicity, and low relative crystallinity (~2.20%), which confer greater flexibility [13]. These modifications allow the starch particles to adsorb at the droplet interface and promote its stability.

The modified starch particles can then be used to produce Pickering emulsions that are capable of encapsulating bioactive compounds in food matrices. Bioactive compounds isolated from natural sources provide many functions in food systems, but several reasons limit their application: (1) incompatibility with the food matrix; (2) rapid degradation during food processing; and (3) vulnerability to digestive activity in the biological system [14]. The Pickering emulsions therefore act as transport systems to deliver these bioactive compounds in the food matrix.

This chapter provides a detailed discussion of two distinct methods for producing Pickering emulsions stabilized by starch particles for encapsulating functional and bioactive compounds in food matrices. The first method involves using OSA-modified canihua starch to produce a Pickering emulsion. The second method describes the procedures for producing a Pickering emulsion with encapsulated curcumin, stabilized by cassava starch modified by a combined method (through HMT and antisolvent precipitation).

2 Materials

2.1 Method 1: Materials Used in the Production of OSA-Modified Canihua Starch and Pickering Emulsion

The native starch of canihua grains was isolated by wet milling in hydroxide solution (0.3 g/100 g), using a semi-industrial blender (Skymsen, Volta Grande, SC, Brazil) [15]. After extraction, the starch was purified by successive washes: three times with the same NaOH solution used previously (0.3 g/100 g) and two times with water. Between each wash, the sample was centrifuged (centrifuge model 5430 R, Eppendorf AG, Hamburg, Germany) at 3000×*g* for 10 min (*see* **Note 1**). The isolated starch had a protein content of about 1 g/100 g and an ash content of 0.6 g/100 g wet basis (w.b.).

The OSA reagent (TRIGON Chemie GmbH, Germany) used was 3 g/100 g of starch for the chemical modification. The emulsion components were soybean oil (density 0.92 g/mL, Sigma-Aldrich, United Sates) as dispersed phase, phosphate buffer (density 1.0096 g/mL; 5 mmol/L, pH 7, 0.2 mol/L NaCl) as continuous phase, and modified canihua starch as stabilizer (*see* **Note 2**). The other reagents and solvents, such as silver nitrate, hydrochloric acid, and phenolphthalein, were of analytical grade.

2.2 Method 2: Materials Used in the Production of Cassava Starch Nanoparticles and Pickering Emulsion with Encapsulated Curcumin

Cassava starch (Siamar, Brazil) was employed as the macromolecule. The ingredients used to prepare the Pickering emulsions included canola oil (Liza/Cargill, Brazil), which was purchased from a local market, and curcumin (*Curcuma longa*) from Sigma Aldrich (United States). Absolute ethyl alcohol (99.5% PA) was supplied by Exodo Cientifica (Brazil), and sodium benzoate (99% PA) was sourced from Sigma Aldrich (United Sates).

3 Methods

3.1 Method 1

The apparatus used for this modification is shown in Fig. 1.

3.1.1 Production of OSA-Modified Canihua Starch

1. Weigh 40 g of the sample of native starch and disperse it in 200 mL of water (20% starch dispersion) with gentle stirring.

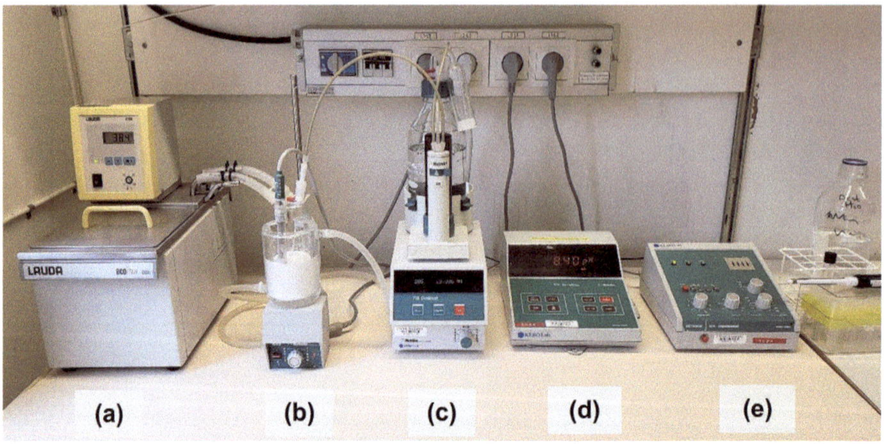

Fig. 1 Chemical starch modification system with OSA: (*a*) water bath with temperature control; (*b*) reactor containing the starch dispersion and pH meter electrode above a magnetic stirrer plate; (*c*) automatic titrator; (*d*) pH meter; (*e*) automatic controller connecting to the titrator with pH meter

2. Hydrate the dispersion of starch granules in a water bath to a temperature of 35 °C for 15 min.
3. Keep the dispersion in the water bath and adjust the pH to 7.9 with NaOH solution (1 mol/L) (*see* **Note 3**).
4. Immediately, add the OSA (divided into four equal parts of 0.3 g each) with gentle stirring every 30 min for each part, maintaining the selected pH (7.9) with NaOH solution (3 mol/L) for a reaction time of 2 h (*see* **Note 4**).
5. After this period, keep the reaction pH (7.9) for three more hours using the NaOH solution (3 mol/L).
6. Maintain gentle stirring of the dispersion in a water bath (35 ° C) until a total reaction time of 24 h.
7. Neutralize the dispersion using HCl solution (1 mol/L) and centrifuge at 10,000×*g* for 20 min at 4 °C.
8. Wash the precipitate twice with water and twice with absolute ethanol in a 1:5 mass/volume ratio.
9. Finally, spread the starch (precipitate) in thin layers onto petri dishes and dry at room temperature (approximately 25 °C) for 24 h.

To obtain the degree of substitution (DS) of the modified starch and reaction efficiency (RE), the next steps were carried out:

1. Weigh 1.5 g of modified starch and add 9 mL of HCl (2.5 mol/ L) then stir gently for 30 min.
2. Centrifuge at 3000×*g* for 10 min at 15 °C.

3. Wash the precipitate with isopropanol solution (90%) under stirring for 10 min (to favor chlorine removal from the dispersion).
4. To evaluate the presence of Cl ions, add one drop of $AgNO_3$ (0.1 mol/L) to the supernatant and evaluate the formation of a white ring on the surface.
5. Repeat the washing process (**step 3**) until there is no reaction with $AgNO_3$.
6. Wash the precipitate with ethanol solution (80%) and dry at room temperature (approximately 25 °C) for 24 h.
7. After 24 h, weigh 100 mg of modified starch and disperse it in 30 mL of water to be subjected to boiling for 30 min in a water bath.
8. Cool the samples to room temperature (approximately 25 °C) and titrate with NaOH (0.01 mol/L) to pH 7 (*see* **Note 5**).

The DS was calculated using Eq. 1:

$$\mathrm{DS} = \frac{0.162 \times (A \times M)/W}{1 - [0.210 \times (A \times M)]/W} \tag{1}$$

where A is the titration volume of NaOH solution (mL), M is the molarity (mol/L) of NaOH solution, 0.162 and 0.210 are the molecular weights (mg/mol) of one anhydrous glucose unit and octenyl succinic anhydride, respectively, and W is the dry weight (g) of the OSA starch. The reaction efficiency (RE) was calculated as follows:

$$\mathrm{RE} = \frac{\text{Actual DS}}{\text{Theoretical DS}} \tag{2}$$

The actual DS was obtained from Eq. 1, and the theoretical DS was calculated by assuming that all the added OSA has reacted with starch and formed the ester derivative. Considering the amount of OSA used in the modification (3% of the starch mass) and the hydroxyl groups available to react in each glucose molecule. In this case, the theoretical maximum degree of substitution was 22×10^{-3} [16] (*see* **Notes 6** and 7).

3.1.2 Production of Pickering Emulsion

To produce a Pickering emulsion with OSA-modified canihua starch, the following steps were performed.

1. Prepare the emulsion in a 1:9 mass ratio (1 g oil/9 g continuous phase (phosphate buffer) and 100 mg modified starch/mL of soybean oil (dispersed phase)). Previously, the modified starch was dispersed in the continuous phase (*see* **Note 8**).
2. Homogenize the emulsions using a rotor-stator high-shear homogenizer (Ystral D-79828, Ballrechtem-Dottingen, Germany).

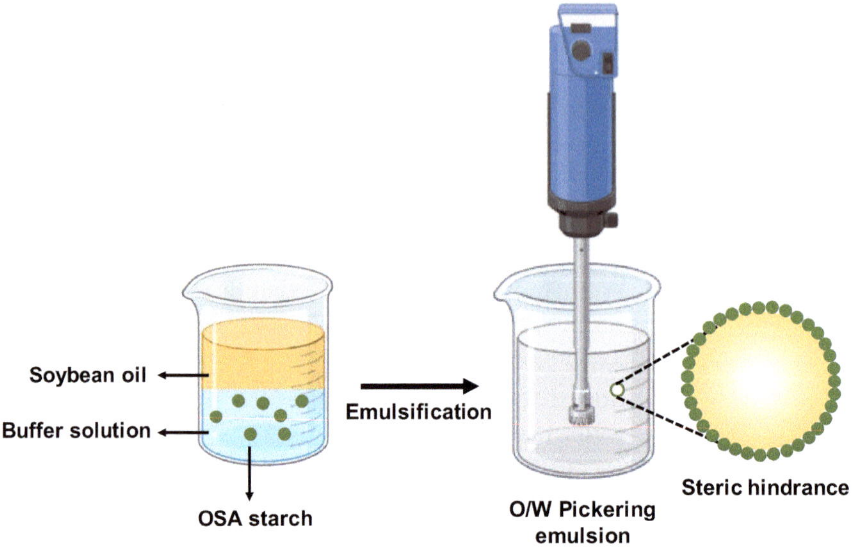

Fig. 2 Production scheme for Pickering emulsion stabilized by OSA starch

3. The conditions of preparation were 6 mm dispersing tool as shear homogenizer, at 22,000 rpm for 30 s [15, 17]. Figure 2 illustrates how the Pickering emulsion was produced.

3.2 Method 2

3.2.1 Production of Cassava Starch Nanoparticles

To produce cassava starch nanoparticles, two sequential modifications of the starch were performed: HMT modification and antisolvent precipitation.

Modification of Cassava Starch by Heat-Moisture Treatment (HMT)

The starch modification with HMT was carried out following the method described by Ramos et al. [13].

1. Adjust the moisture content wet basis (w.b.) of the cassava starch to 20% by adding distilled water and thoroughly homogenizing the sample (*see* **Note 9**).
2. Store the starch in airtight containers under refrigeration (~8 °C) for 24 h to allow moisture equilibration.
3. After this period, spread the starch in thin layers onto petri dishes, covered with two layers of plastic film.
4. Place the dishes with the starch in an air convection oven (Tecnal, TE-394/3, Piracicaba, SP, Brazil) at 130 °C for 4 h (*see* **Note 10**).
5. Allow the samples to cool on the bench to room temperature (approximately 25 °C).
6. Grind the starch using a mortar and pestle, if necessary.

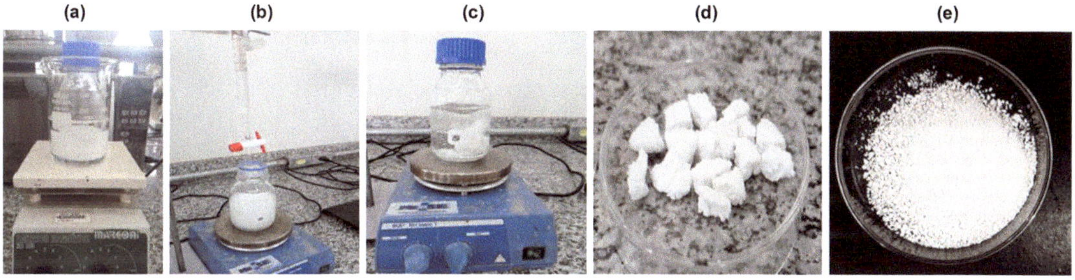

Fig. 3 Steps in the production of cassava starch nanoparticles: (**a**) gelatinization of the starch in a water bath and heating plate; (**b**) precipitation with ethanol drop by drop; (**c**) stirring of the particles for 12 h; (**d**) cutting of the particle mass before freezing and lyophilization; (**e**) starch nanoparticles after freeze-drying, grinding, and sieving

7. Sieve the starch using a 100-mesh sieve (0.149 mm) with the help of a spatula and store it in airtight containers.

Nanoprecipitation of Cassava Starch with Antisolvent

For starch precipitation by antisolvent, the method described by Ramos et al. [13] was employed.

1. In a flask with a sealing cap, weigh 5 g of HMT-modified cassava starch (dry basis, d.b.) and add ultrapure water to reach a total weight of 100 g.
2. Seal the flask and gelatinize the dispersion in a water bath at 95 °C for 30 min, using a magnetic stirrer plate at 600 rpm (IKA, C-MAG HS7, Labotechnik, Staufen, Germany) (Fig. 3a) (*see* **Note 11**).
3. After this period, keep the gel under magnetic stirring until it cools down (approximately 25 °C).
4. Prepare a burette with 100 mL of absolute ethanol.
5. Position the open flask containing the starch gel directly below the burette with ethanol, ensuring continuous stirring.
6. Slowly add the 100 mL of absolute ethanol to the starch gel flask (approximately 1 drop/s) (Fig. 3b) (*see* **Note 12**).
7. Let the mixture stir for 12 h at 25 °C, 600 rpm (Fig. 3c).
8. Then, remove the nanoparticle mass from the flask using tweezers and transfer it to a 50-mL Eppendorf tube. Discard the remaining liquid in the gel flask (*see* **Note 13**).
9. Add approximately 30 mL of absolute ethanol to the Eppendorf tube containing the sample.
10. Centrifuge (model 5430 R, Eppendorf AG, Hamburg, Germany) the sample at 3000×*g* for 15 min.
11. Discard the liquid and keep the nanoparticle mass in the Eppendorf tube.

12. Wash the starch nanoparticle two more times with absolute ethanol, repeating **steps 9**, **10**, and **11**.
13. Using tweezers, remove the starch nanoparticle from the tubes and dry it in an oven, in petri dishes, for 10 min at 60 °C to remove excess ethanol.
14. Cool the sample to room temperature (approximately 25 °C).
15. Cut the sample into smaller pieces using spatulas (*see* **Note 14**) (Fig. 3d).
16. Freeze the sample for 48 h in petri dishes with lids, wrapped in plastic film, in a freezer (approximately −18 °C) (*see* **Note 15**).
17. Lyophilize the sample (LC5500, Terroni, São Carlos, SP, Brazil) for 48 h.
18. Grind the sample using a mortar and pestle.
19. Sieve the starch nanoparticles (SNP) in a 70-mesh sieve (0.210 mm) and store them in airtight containers (Fig. 3e).

3.2.2 Production of Pickering Emulsion with Encapsulated Curcumin

The production of oil in water (O/W) Pickering emulsions followed the method of Ramos et al. [18]. The formulation detailed here was the one that showed the highest stability in our studies and also included encapsulated curcumin [18].

1. To produce 20 g of emulsion, weigh 0.8 g of SNP into a 50-mL glass beaker, along with 15.19 g of distilled water and 0.004 g of sodium benzoate (*see* **Note 16**).

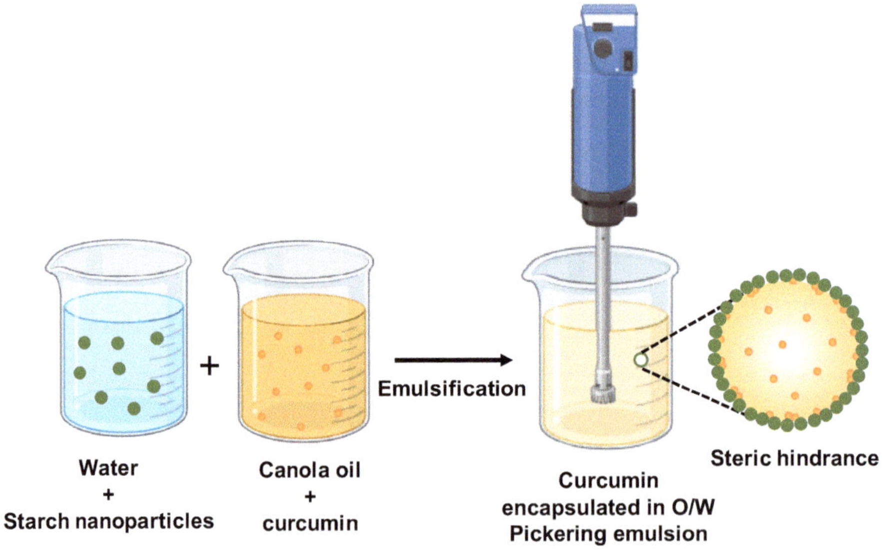

Fig. 4 Production scheme for Pickering emulsion with encapsulated curcumin

2. In a separate beaker of about 30 mL, weigh 3.99 g of canola oil and 0.014 g of curcumin (*see* **Notes 17** and **18**). Figure 4 illustrates how the Pickering emulsion was produced.
3. Place the beaker with the oil and curcumin under magnetic stirring for 10 min at 300 rpm.
4. Combine the oil with curcumin into the beaker containing the aqueous phase.
5. Emulsify the mixture using a rotor-stator homogenizer (IKA, T25, Labotechnik, Staufen, Germany) at 14,000 rpm for 3 min.

4 Notes

1. In the native starch purification stage, the brown surface layer, associated with proteins, that formed in the sediment between each centrifugation process was manually removed.
2. In the production of the Pickering emulsion, a phosphate buffer was used to avoid possible changes in the pH of the emulsion.
3. In the chemical modification stage with OSA, there are pH modifications when adding the reagent; ensure that these modifications are minimal or quickly corrected.
4. In the chemical modification stage, the incorporation of OSA is carried out in four stages to avoid saturation of the medium and to guarantee maximum incorporation of OSA groups into the starch structure.
5. When determining DS, an agitator might be necessary in the beginning to ensure a homogeneous solution while titrating.
6. The minimum DS required can vary depending on several factors, such as the type of starch used, the modification conditions, and the desired emulsion characteristics (e.g., droplet size, stability, etc.). Generally, to stabilize Pickering emulsions with OSA-modified starch, DS typically ranges from 0.64×10^{-2} to 3.1×10^{-2} [16, 19], which is sufficient to impart amphiphilic properties to the modified starch. The DS value of canihua OSA starch with the best emulsifying capacity obtained in our laboratory was approximately 0.018.
7. This modification method can be used for different sources of starch [20–22].
8. Pickering emulsions stabilized with OSA starch can be used to encapsulate hydrophilic or/and hydrophobic bioactive compounds. In any case, before encapsulation, it is necessary to dissolve the active compound in the phase to be encapsulated [23–25].

9. For a total of 200 g of cassava starch, initially at 12.84% moisture content (w.b.), 17.90 mL of distilled water were added to achieve 20% moisture content (w.b.).
10. If any plastic wrap on the petri dishes with starch punctures or tears during heating in the oven, the starch should be discarded, since the plastic hinders moisture loss and avoids excessive drying or burning.
11. The water bath can be done with a larger beaker (~500 mL) in direct contact with the heating plate, containing water and the flask with the lid and the starch suspension inside.
12. HMT-modified cassava starch tends to form a mass of nanoparticles (ball) after the addition of ethanol is complete. This phenomenon does not interfere with the production of the nanoparticles.
13. The liquid in the flask can be discarded if a nanoparticle mass has formed, which can be easily removed with tweezers. If no mass forms, the entire liquid in the flask should be centrifuged according to **step 10**.
14. Breaking the nanoparticle mass into smaller pieces facilitates the complete lyophilization of the sample interior.
15. The freezing of SNP can also be done with liquid nitrogen by immersing the sample for a few minutes until it is completely frozen.
16. The Pickering emulsion formulation detailed here was 80 g/100 g aqueous phase (including distilled water +4 g/100 g SNP + 0.02 g/100 g sodium benzoate) and 20 g/100 g oil phase (including oil +0.07 g/100 g curcumin).
17. The maximum amount of the bioactive curcumin is related to its solubility in the phase it was added (canola oil).
18. Other types of lipophilic bioactive compounds could be added in the same way to the oil phase of this Pickering emulsion system.

Acknowledgments

Part of this research was funded by FONDECYT-CONCYTEC/Peru (grant contract number 240-2015-FONDECYT) and CAPES—Brazil (Sandwich Doctorate Program Abroad, process no. 88881.190627/2018-01).

This research was funded by "Coordenação de Aperfeiçoamento de Pessoal de Nível Superior—Brasil (CAPES)"—Finance Code 001.

References

1. Ming L, Wu H, Liu A et al (2023) Evolution and critical roles of particle properties in Pickering emulsion: a review. J Mol Liq 388: 122775
2. Cheon J, Haji F, Baek J et al (2023) Pickering emulsions for functional food systems. J Agric Food Res 11. https://doi.org/10.1016/j.jafr.2023.100510
3. Du M, Chen L, Din Z et al (2023) Structure and surface properties of ozone-conjugated octenyl succinic anhydride modified waxy rice starch: towards high-stable Pickering emulsion. Int J Biol Macromol 253:126895. https://doi.org/10.1016/j.ijbiomac.2023.126895
4. Shahbazi M, Jager H, Ettelaie R et al (2024) A smart thermoresponsive macroporous 4D structure by 4D printing of Pickering-high internal phase emulsions stabilized by plasma-functionalized starch nanomaterials for a possible delivery system. Curr Res Food Sci 8: 100686. https://doi.org/10.1016/j.crfs.2024.100686
5. Lin X, Li S, Yin J et al (2020) Anthocyanin-loaded double Pickering emulsion stabilized by octenylsuccinate quinoa starch: preparation, stability and in vitro gastrointestinal digestion. Int J Biol Macromol 152:1233–1241. https://doi.org/10.1016/j.ijbiomac.2019.10.220
6. Zhu F (2019) Starch based Pickering emulsions: fabrication, properties, and applications. Trends Food Sci Technol 85:129–137
7. Ramirez-Lopez S, Ditchfield C, Moraes ICF (2023) Physicochemical properties of grain and starch from kanihua (chenopodium pallidicaule) compared with quinoa (chenopodium quinoa) Originated from Peru. Chem Eng Trans 102:49–54. https://doi.org/10.3303/CET23102009
8. Zhu F (2015) Composition, structure, physicochemical properties, and modifications of cassava starch. Carbohydr Polym 122:456–480
9. Chen H, Jiang Y, Zhang B et al (2023) Application of Pickering emulsions stabilized by corn, potato and pea starch nanoparticles: effect of environmental conditions and approach for curcumin release. Int J Biol Macromol 238:124115. https://doi.org/10.1016/j.ijbiomac.2023.124115
10. Ramos GVC, Suzigan AR, Pinho SC, Moraes ICF (2023) Impact of emulsification time and concentration of modified starch nanoparticles on Pickering stability. Chem Eng Trans 102: 247–252. https://doi.org/10.3303/CET23102042
11. Zhang L, Chen DL, Wang XF et al (2022) Enzymatically modified quinoa starch based Pickering emulsion as carrier for curcumin: rheological properties, protection effect and in vitro digestion study. Food Biosci 49: 101933. https://doi.org/10.1016/j.fbio.2022.101933
12. Sweedman MC, Tizzotti MJ, Schäfer C, Gilbert RG (2013) Structure and physicochemical properties of octenyl succinic anhydride modified starches: a review. Carbohydr Polym 92: 905–920
13. Ramos GVC, Rabelo MEA, de Pinho SC et al (2024) Dual modification of cassava starch using physical treatments for production of Pickering stabilizers. Food Secur 13. https://doi.org/10.3390/foods13020327
14. Hu YT, Ting Y, Hu JY, Hsieh SC (2017) Techniques and methods to study functional characteristics of emulsion systems. J Food Drug Anal 25:16–26
15. Rayner M, Sjöö M, Timgren A, Dejmek P (2012) Quinoa starch granules as stabilizing particles for production of Pickering emulsions. Faraday Discuss 158:139–155. https://doi.org/10.1039/c2fd20038d
16. Song X, He G, Ruan H, Chen Q (2006) Preparation and properties of octenyl succinic anhydride modified early indica rice starch. Starch/Staerke 58:109–117. https://doi.org/10.1002/star.200500444
17. Marefati A, Bertrand M, Sjöö M et al (2017) Storage and digestion stability of encapsulated curcumin in emulsions based on starch granule Pickering stabilization. Food Hydrocoll 63: 309–320. https://doi.org/10.1016/j.foodhyd.2016.08.043
18. Ramos GVC, de Pinho SC, Gomes A et al (2024) Designing Pickering emulsions stabilized by modified cassava starch nanoparticles: effect of curcumin encapsulation. Processes 12. https://doi.org/10.3390/pr12071348
19. Wang J, Su L, Wang S (2010) Physicochemical properties of octenyl succinic anhydride-modified potato starch with different degrees of substitution. J Sci Food Agric 90:424–429. https://doi.org/10.1002/jsfa.3832

20. Zhao Y, Khalid N, Shu G et al (2017) Formulation and characterization of O/W emulsions stabilized using octenyl succinic anhydride modified kudzu starch. Carbohydr Polym 176:91–98. https://doi.org/10.1016/j.carbpol.2017.08.064
21. Miao M, Li R, Jiang B et al (2014) Structure and physicochemical properties of octenyl succinic esters of sugary maize soluble starch and waxy maize starch. Food Chem 151:154–160. https://doi.org/10.1016/j.foodchem.2013.11.043
22. Timgren A, Rayner M, Dejmek P et al (2013) Emulsion stabilizing capacity of intact starch granules modified by heat treatment or octenyl succinic anhydride. Food Sci Nutr 1:157–171. https://doi.org/10.1002/fsn3.17
23. Matos M, Timgren A, Sjöö M et al (2013) Preparation and encapsulation properties of double Pickering emulsions stabilized by quinoa starch granules. Colloids Surf A Physicochem Eng Asp 423:147–153. https://doi.org/10.1016/j.colsurfa.2013.01.060
24. Marefati A, Sjöö M, Timgren A et al (2015) Fabrication of encapsulated oil powders from starch granule stabilized W/O/W Pickering emulsions by freeze-drying. Food Hydrocoll 51:261–271. https://doi.org/10.1016/j.foodhyd.2015.04.022
25. Zhu F (2017) Encapsulation and delivery of food ingredients using starch based systems. Food Chem 229:542–552

Chapter 11

Encapsulation of Food Bioactive Compounds Using PDMS-Based Microfluidic Devices

Cristhian Rafael Lopes Francisco, Tatiana Porto Santos, Mariano Michelon, and Rosiane Lopes da Cunha

Abstract

Microfluidics is a low-cost and easy-operation technique that can be used to design efficient delivery systems with well-controlled features (e.g., size and morphology). Herein, the basic aspects of microfluidics and the advantages are described, as well as what a beginning researcher needs to know before starting experiments using this technique. In this chapter, simple step-by-step guidance is provided on how the formulation of the delivery system, the construction of the microchip, and its operation can affect the features of the main structures designed from templates based on water-in-oil or oil-in-water emulsions, in addition to liposomes. Details are provided on how to produce and operate PDMS-based devices for a successful experimental design.

Key words Emulsion, Flow-focusing, Gelation, Liposome, Microgel

1 Introduction

Microfluidics is defined as the science and technology of systems that process or manipulate small amounts (10^{-9} to 10^{-18} l) of fluids using channels with dimensions of tens to hundreds of micrometers [1]. By dealing with such small samples, microfluidic devices (i.e., a chip containing a channel of at least one-micrometer dimension) have gained significant attention as an interesting tool for designing delivery systems (DS) with specific features, mainly vehiculating materials that are difficult to obtain and have high added value [2]. This is possible because the small dimension of the channels promotes a laminar flow (i.e., fluid layers organized flowing with no adjacent mixing) that allows a controlled environment for structure formation. Additionally, the high area/volume ratio of microchannels results in enhanced mass and heat transfer. Microchannels can be operated with a monophasic laminar flow to

Samantha Cristina de Pinho and Carmen S.F. Trindade (eds.), *Bioactives Encapsulation: Food Applications*, Methods and Protocols in Food Science, https://doi.org/10.1007/978-1-0716-4538-3_11,

produce, e.g., micro- and nanoparticles, complexes, and liposomes, or in a multiphase system in which droplets are formed (i.e., droplet microfluidics) to assemble emulsion-based vehicles [3–9].

Since DS can be easily produced by bulk techniques, what are the advantages of microfluidics? Microfluidics stands out from more common encapsulation techniques due to its low energy demand, ability to handle small amounts of reagents, and the possibility of easily tuning the DS properties [10]. Furthermore, the same device can be used to produce a variety of both mono- and multiphase encapsulation systems depending on the properties of the material used for its construction or simply by modifying the surface properties of the channel wall [11, 12]. All these characteristics together result in a low-cost technique capable of producing DS with precise control of structure size and chemical composition (e.g., by varying the flow rate ratio of stream(s)). However, the throughput of DS production with microfluidics is small, and therefore parallelization has been investigated to improve and disseminate the use of microfluidics [13–15].

The control of size, polydispersity, and composition is especially crucial in the production of DS since tuning these features allows modulating the physical stability of the structure, the chemical stability of the bioactive compound (BC), as well as its release rate [16, 17]. This precise tuning can be achieved in microfluidics by correctly setting the physical parameters that affect structure formation within the microdevice, such as the chip design, construction material, surface properties, and flow regime of the stream (s) [18].

Microchannels can be produced with different geometries (designs), which will dictate the fluid dynamics involved in the formation of the structure. Shear-based microfluidic devices, such as planar and 3D glass capillary devices, are commonly used to produce multi- and monophase DS, such as emulsions and nanoparticles [4, 19, 20], while spontaneous formation-based devices are restricted to producing emulsions and bubbles [21, 22]. Considering the different designs of microchannels, flow-focusing has proven to be a versatile design to generate not only droplets in the microscale range but also nanoparticles (such as liposomes) with controlled size and low polydispersity. Several studies have shown that the size, polydispersity, and lipid composition of liposomes can be easily tuned by changing the flow rates of the organic and aqueous solutions using this design [23].

Knowledge of material science is especially an advantage when dealing with the flow of liquid solutions within small channels where wettability plays an important role. Microchannels can be built with a variety of materials, although for microfluidic systems that deal with the delivery of food-grade components, most studies focus on the use of glass and polydimethylsiloxane (PDMS) [24]. PDMS is the most used material for microfluidic studies related to food science due to its transparency, flexibility, and low toxicity, being also simple to implement and inexpensive [1]. However, its use for application in food products is still limited due to swelling and degradation when in contact with some organic solvents, acids, or bases [25], as well as nonspecific adsorption and permeation by hydrophobic molecules [26, 27].

PDMS-based chips are naturally hydrophobic and therefore suitable to produce water-in-oil emulsions (encapsulating hydrophilic bioactive compounds (HBC)) and aqueous-based systems where surface wettability is not mandatory, such as liposomes (encapsulating both HBC and lipophilic bioactive compounds (LBC)). Alternatively, PDMS can be easily hydrophilized by applying different treatments, e.g., oxygen plasma, allowing the production of oil-in-water emulsions (encapsulating LBC) [28]. Additionally, surface coating of PDMS with food-grade polyelectrolytes is suggested as a promising alternative to enable the use of PDMS devices to produce foods in the future [24]. Still, the use of PDMS devices for prospecting studies in the design of food-grade DS seems the most reasonable, due to the several advantages previously described. Therefore, due to the relevance of PDMS in the context of microfluidics, focus will be given to this material in the current protocol.

Finally, once having a microfluidic device that meets all the requirements to produce a specific DS, the operator must know how to organize the microfluidic setup. Microfluidic setups are commonly composed of a flow controller responsible for inserting the stream(s) into the device through tubes, a microscope to visualize the formation of the structure, and a collector. Another important aspect to be considered is the establishment of the regime of structure formation that will strongly affect the features of the final structure. Dynamic flow parameters such as stream(s) flow rate and stream(s) flow rate ratio can, for instance, affect the droplet formation regime in emulsions and diffusional mass transfer in liposome production. Therefore, systematic preliminary studies are important to find specific operating conditions to tune the features of the DS to be produced. Furthermore, depending on the DS produced, further purification steps may be necessary to obtain a product with no residual solvents. Paying attention to all these details will be crucial to the success of the experiment of a beginner researcher in the context of encapsulating BC using microfluidics.

This chapter was designed in a way young researchers curious about microfluidics would find a guide to start their experiments on model DS encapsulating HBC and LPC using microchannels. To make learning more concise and precise, we will focus on how to produce (1) a PDMS-based microfluidic device, (2) water-in-oil emulsion template microgels to encapsulate HBC, (3) oil-in-water emulsions to encapsulate LBC, and (4) liposomes to encapsulate LBC. Valuable notes and references will be provided to discuss in detail all the necessary fundamentals regarding device design and construction, sample preparation, flowing parameters adjustment, and purification of the delivery system.

2 Material

2.1 Microchip Fabrication

1. Spin rinse dryer.
2. Spin coater.
3. Mask aligner with UV exposure.
4. Hot plate.
5. Cleaning solution (e.g., Nano-Strip®).
6. Sylgard 184 Silicone Elastomer kit containing the polydimethylsiloxane pre-polymer and curing agent.
7. SU-8 UV photoresist and SU-8 developer solution.
8. Silicon wafer.
9. Glass slides.
10. Tubing (e.g., polyetheretherketone (PEEK), polytetrafluoroethylene (PTFE), Tygon®, fluorinated ethylene propylene (FEP), ethylene tetrafluoroethylene (ETFE), and polypropylene (PP)).
11. Silicone adhesive glue.
12. Syringe needles.

2.2 Microchip Operation

1. Flow controller (e.g., syringe pumps or pressure-based flow controllers).
2. An optical microscope coupled with a camera and imaging software.

2.3 Water-in-Oil Emulsion Template Microgels to Vehiculate HBC

1. Food-grade oil (e.g., long-chain triglycerides (LCTs) such as soybean, sunflower seed, and corn).
2. Distilled water, ultrapure water, or buffer (if necessary).
3. Cold set gel-forming agent (e.g., proteins (soy and whey protein (WPI)) and polysaccharides (gellan gum, sodium alginate, or low methoxylated pectin)).
4. Low hydrophilic-lipophilic balance (HLB) (3–8) surfactant or natural emulsifier capable of forming water-in-oil emulsions (e.g., polyglycerol polyricinoleate (PGPR), sorbitan laurate (Span 20), sorbitan palmitate (Span 40), Span stearate (Span 60), sorbitan oleate (Span 80), sorbitan trioleate (Span 85), modified polysaccharide, and protein-based particles [29]).
5. Crosslinking agent (e.g., mono- and divalent salts and enzymes).
6. Hydrophilic bioactive compound to be encapsulated (e.g., phenolic compounds, glucosinolates, and vitamins).

2.4 Oil-in-Water Emulsion Loaded to Vehiculate LBC

1. Food-grade oil (e.g., long-chain triglycerides (LCTs) such as soybean, sunflower seed, and corn).
2. Distilled water, ultrapure water, or buffer (if necessary).
3. High HLB (8–18) surfactant or natural emulsifier capable of forming oil-in-water emulsions (e.g., polyoxyethylene sorbitan laurate (Tween 20), polyoxyethylene sorbitan palmitate (Tween 40), polyoxyethylene sorbitan stearate (Tween 60), polyoxyethylene sorbitan oleate (Tween 80), polyoxyethylene sorbitan trioleate (Tween 85), animal (whey, sodium caseinate, egg albumin), and plant (soy, pea, lentil) proteins).
4. LPC to be encapsulated (e.g., phytosterols, terpenoids, carotenoids, tocopherols, triterpenes, capsaicinoids, polyunsaturated fatty acids (PUFAs), and vitamins).

2.5 Liposomes to Vehiculate LBC

1. Absolute ethanol.
2. Distilled water, ultrapure water, or buffer (if necessary).
3. Phospholipids (e.g., soy lecithin, egg lecithin, marine lecithin, milk phospholipids, dipalmitoylphosphatidylcholine (DPPC), distearoylphosphatidylcholine (DSPC), and dimyristoylphosphatidylcholine (DMPC)).
4. LPC to be encapsulated (e.g., phytosterols, terpenoids, carotenoids, tocopherols, triterpenes, capsaicinoids, polyunsaturated fatty acids (PUFAs)).

3 Methods

3.1 Master Mold Production

Steps 2–6 must be performed in a clean room to avoid contamination.

1. Create a flow-focusing microchannel design using CAD software and print it in high resolution onto a transparent film using a commercial printer with the design clear and the background ink. This will be called a "photomask" (*see* **Note 1**) (Fig. 1a).
2. Soak the silicon wafer in a cleaning solution (e.g., Nano-Strip®) for 10 min and rinse well with deionized water.
3. Dry the wafer using a spin rinse dryer and place it on a hot plate at 200 °C for 20 min.
4. Spin-coat SU-8 photoresist onto the silicon wafer (*see* **Note 2**).
5. Apply UV light to the photomask-covered wafer using a contact mask aligner (*see* **Note 3**).
6. Place the wafer into SU-8 developer for the time recommended by the supplier and then rinse with isopropyl alcohol.

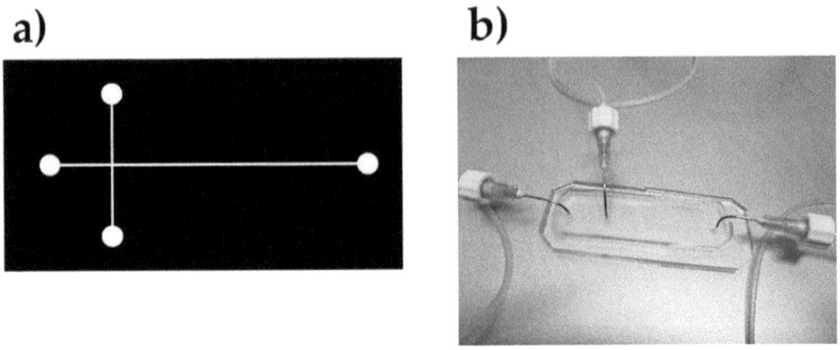

Fig. 1 An example of a photomask with a flow-focusing design (**a**) and an image of a fabricated microfluidic device (**b**)

3.2 PDMS Device Replication

1. Place the mold on a resistant foil sheet and fold it up around the sides of the mold to create a lip (*see* **Note 4**).
2. Mix Sylgard 184 silicon elastomer base and curing agent in a 10:1 (w/w) ratio (*see* **Note 5**).
3. Degas the mixture using a vacuum pump for 40 min or until the bubbles are completely eliminated (*see* **Note 6**).
4. Pour the mixture into the mold and transfer it to an oven at 65 °C for 1 h.
5. Remove the mold from the oven and wait for it to cool down. Peel off the PDMS from the mold, trim the edges of the microchannel, and pierce holes into the inlet(s) and outlet (s) marks (*see* **Note 7**).
6. Cover the microchannel with magic tape to prevent dust deposition and store it until use.

3.3 Sealing the Microchannel

1. Treat with oxygen plasma the PDMS microchannel along with a glass slide or a PDMS slab (*see* **Notes 8** and **9**).
2. Bring the PDMS microchannel and the glass slide/PDMS slab together and gently press them against each other.
3. Check successful sealing by gently pulling the edges of the PDMS slabs (*see* **Note 10**).
4. Wait at least 24 h to allow the PDMS chip to return to its natural hydrophobic character (*see* **Note 11**).
5. The same microfluidic device can be used to produce emulsions or liposomes (*see* **Note 1**).

3.4 Production of Water-in-Oil Emulsion Template Microgels Loaded with HBC

3.4.1 Production of Dispersed and Continuous Phases

1. Dispersed phase: Prepare an aqueous dispersion of the gelling agent at the desired concentration using magnetic stirring for adequate time and temperature for complete dissolution (if necessary). Afterward, add the HBC to the aqueous dispersion at room temperature (*see* **Notes 12** and **13**).
2. Continuous phase: Add the surfactant/emulsifier and the crosslinking agent to the oil in the concentration necessary to

stabilize the droplets. Stir the mixture at room temperature and then immerse it in an ultrasonic bath for 30 minutes (*see* **Notes 14** and **15**).

3.4.2 Production of Microgels

1. Transfer the dispersed and continuous phases to the vials of the flow controller and assemble the tubing and fittings (*see* **Note 16**).
2. Connect the filled vessels to the chip using tubes connected to the syringe needles. Insert the syringe needles carefully into the chip inlets. Insert a syringe needle connected to a tube into the microchannel outlet to direct the formed microgels into a collection vessel. Apply silicone adhesive glue where the tubes were connected to the chip to prevent leakage (Fig. 1b). If the tube is rigid enough, it can be inserted directly into the chip without using connectors/syringes.
3. Place the chip onto the microscope holder and secure it with adhesive tape.
4. Turn on the microscope and camera. Adjust the lens to clearly view channels. Check that the channels are intact, with no deformation or clothing. Focus the camera on the conjunction of channels.
5. Start the flow controller with a low flow rate (<10 μL/min) for both phases and wait for them to reach the junction channel.
6. Set the disperse phase to a fixed flow rate value and begin increasing the flow rate of the continuous phase until an intermittent dripping flow regime is achieved (Fig. 2a) (*see* **Notes 17–19**).
7. Wait for the microgel phase to separate in the collection vessel and collect it. Remove the oil and store the microgels at 4 °C to complete gelation (*see* **Note 20**).
8. Characterize the microgels in terms of size distribution, coefficient of variance, and visual appearance (*see* **Note 21**).

3.5 Production of Oil-in-Water Emulsion Loaded with LBC

3.5.1 Production of Dispersed and Continuous Phases

1. Disperse phase: Prepare a dispersion of LBC with oil at the desired concentration using magnetic stirring for adequate time and temperature for complete solubilization (if necessary) (*see* **Note 22**).
2. Continuous phase: Add the surfactant/emulsifier to the water/buffer at the concentration necessary to stabilize the droplets. Stir the mixture at room temperature until complete solubilization.

3.5.2 Production of Emulsions

1. Immediately after **step 3** of Subheading 3.3, flood the entire microchannel with the continuous phase (i.e., aqueous phase) to maintain the hydrophilicity of the chip. After that, emulsion production should be started as soon as possible.

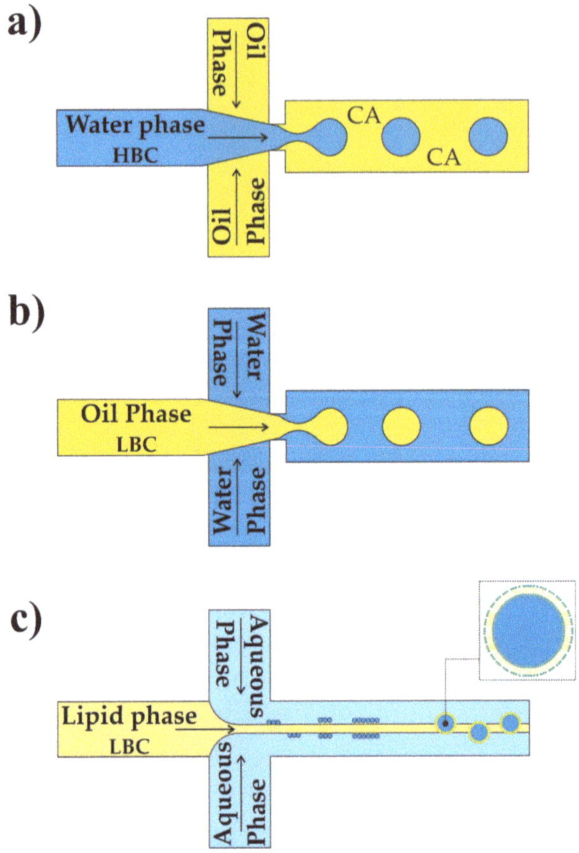

Fig. 2 Flow-focusing devices to produce (**a**) water-in-oil emulsions template microgels, (**b**) oil-in-water emulsions, and (**c**) liposomes. Note: CA stands for crosslinking agent. (Figures are not drawn to scale)

2. Follow **steps 1–6** described in Subheading 3.4.2 (Fig. 2b) (*see* **Note 23**).
3. Collect the material and wait for the phases to separate; collect the upper phase and characterize the emulsion as necessary (*see* **Note 24**).

3.6 Production of Liposomes Loaded with LBC

3.6.1 Production of Aqueous and Lipid Phases

1. Aqueous phase: use distilled water, ultrapure water, or buffer.
2. Lipid phase: disperse the phospholipid(s) and the LBC in ethanol under stirring (*see* **Note 25**).

3.6.2 Production of Liposomes

1. Follow **steps 1–5** described in Subheading 3.4.2 now referring to the aqueous and lipid phases instead of the continuous and disperse phases.

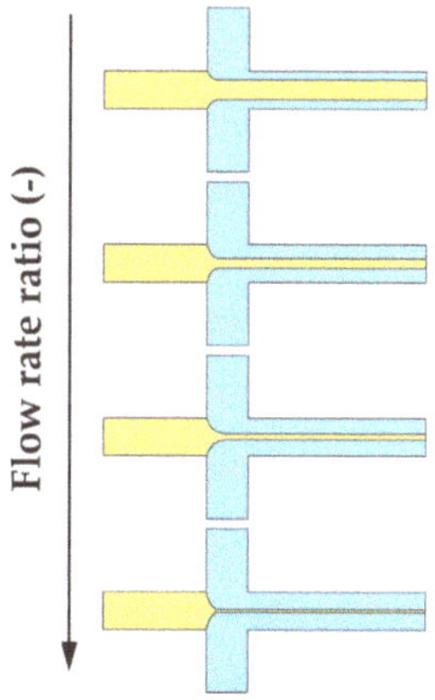

Fig. 3 Representation of the effect of flow rate ratio on the width of the lipid phase. (Note: Yellow and blue stand for the lipid and aqueous phases, respectively. Figures are not drawn to scale)

2. Vary the flow rate ratios of phases and analyze the particle size distribution of different batches (*see* **Notes 26–28**).
3. Produce the liposomes under adjusted operating conditions to the required quantity of particles before removing the ethanol (*see* **Notes 29** and **30**) (Fig. 3).

4 Conclusions

Microfluidics has emerged as an innovative technique to produce structures with tuned features to ensure the target delivery of food bioactive compounds. This step-by-step protocol gathers all the basic knowledge a beginner researcher must have to start experiments on encapsulating hydrophilic and lipophilic bioactive compounds through microfluidics using multi- (water-in-oil and oil-in-water emulsions) and monophase (liposome) delivery systems as models. The valuable notes on the production of the microfluidic device, formulation of the delivery system, and the device operation are currently applied in real laboratories and must be taken as the base for beginner researchers. However, one must be aware that those notes are not exempt from possible modifications. Examples of materials and methods have been provided, but depending on the bioactive compound and other materials used, further adaptations may be necessary. Thus, we encourage the reader to carry out preliminary studies with these guidelines and search for additional literature, taking into account possible adaptations.

5 Notes

1. The layout of the microchannel plays an important role in the droplet formation regime and liposome production. Flow-focusing offers the advantage of promoting shear on both sides of the dispersed phase stream, which helps to "break" the droplet more easily, allowing the production of smaller droplets [30] (Fig. 2a, b). Flow-focusing is also effective in promoting parallel laminar flow of the lipid and aqueous phases, enabling the controlled diffusion of molecules between streams and self-assembly of liposomes (Fig. 2c). The aspect ratio of flow-focusing devices can also affect liposome formation. For more information, check out the previous study performed by [31]. Due to the versatility of this geometry, we will be focusing on it.
2. The channel depth is determined by the thickness of the SU-8 photoresist obtained by spin-coating. For detailed information on which velocity to apply when using each type of SU-8 photoresist, consult the product datasheet.
3. For detailed information on power value and time to be applied, consult the equipment supplier.
4. The lip should be high enough to prevent PDMS from overflowing. If resistant foil is not available, use two or more sheets of regular foil to form a mechanically strong layer.
5. Start stirring the mixture slowly to avoid losing the curing agent, which is a volatile compound. Stir the mixture until it is opaque and full of bubbles; this will ensure the material is well mixed.
6. Centrifuging the PDMS mixture before applying the vacuum drastically reduces the presence of bubbles and thus reduces the suggested vacuum time. Apply vacuum until bubbles are no longer visible.
7. Use a well-sharpened instrument (e.g., a scalpel blade) to easily trim the excess PDMS to avoid the deformation of the microchannel. Trim the microchannel into a rectangular slab, leaving at least a 0.5 cm wide edge from the microchannel. To pierce the inlet(s) and outlet(s) of the microchannel, use instruments that leave holes with smooth surfaces (e.g., a biopsy punch).
8. Oxygen plasma treatment makes PDMS hydrophilic, which allows it to permanently seal when in contact with glass or other PDMS surfaces [32]. Make sure the surface the microchannel is printed on is facing up. Being aware of this is essential to avoid mistakes, such as treating the opposite face of the microchannel slab.

9. Alternatively, PDMS devices can be sealed using the gradient technique in which the PDMS device replica and the slab of PDMS are produced with different concentrations of curing agent, allowing the bonding of these two parts throughout the curing process without the use of additional treatment [33].
10. Successful sealing is essential to prevent leakage during chip operation. If the edges are slightly loose, leave the chip on a hot plate for a few minutes while applying pressure over the chip, and check again for a complete seal.
11. Oxygen plasma treatment of PDMS makes it temporarily hydrophilic. Insufficient waiting time may lead to a partially hydrophobic microchannel, which impairs its correct operation [34].
12. Some HBCs, such as phenolic compounds, can promote gel formation of polysaccharides at specific concentrations. To check this, it is important to perform preliminary studies to determine the concentrations of the gelling agent and the phenolic compound that do not form a gel when they are mixed. This knowledge is essential to produce liquid mixtures capable of being pumped.
13. Further preliminary testing is needed to identify the concentration of gelling agent that can form a gel when in contact with the crosslinking agent. This is essential to determine the formulation capable of forming microgel inside the chip.
14. Perform preliminary tests to determine the best surfactant concentration to form stable water-in-oil emulsions and the best crosslinking agent concentration to form a gel. Satisfactory preliminary results can be obtained using bulk methods [9]. The choice of the crosslinking agent must consider that the gel will not be formed during the formation of the droplet but during the residence time in the channel.
15. Microgels can be formed from a water-in-oil emulsion by internal or external gelation. In the case of external gelation, the crosslinking agent must be at least partially soluble in the oil (continuous phase), such as calcium acetate. In this scenario, calcium diffusion through the microgels may be delayed due to the formation of a thick-gelled layer at the interface at the beginning of the gelation process, requiring more time to form a solid structure. Internal gelation, on the other hand, can produce microgels with a more homogeneous structure. In this case, e.g., a water-insoluble calcium salt is added to the dispersed phase (water containing the gelling agent-BC mixture). The oil is acidified to induce the release of Ca^{2+} from the insoluble salt to crosslink the gelling agent. For more information on this topic, we recommend the review papers produced by Chan et al. [35] and Chen et al. [36].

16. The volume of the vessel should be large enough to produce the required amount of product without resetting the experiment to refill the vessels. Stopping the operation will disturb the steady-state flow.
17. Droplets can be formed basically through three different regimes named squeezing, jetting, and dripping. Of these three, dripping is associated with the formation of more uniform droplets. In this regime, the droplet detaches at the tip of the conjunction channel before completely occupying the cross-section area of the channel (Fig. 2a). For more information, see the study performed by Kovalchuk and co-workers [37].
18. Droplet size and thus microgel size can be easily tuned by changing the flow rate ratio of dispersed and continuous phases. Droplet size and polydispersity are less affected by changes in total flow than by the flow rate ratio of phases. Therefore, the researcher is encouraged to play with these parameters to produce monodisperse microgels with high throughput.
19. Once desired flow rates are achieved, allow the system to stabilize for a few minutes (reaching a steady-state regime). Syringe pumps may take longer to stabilize flow compared to pressure-based flow controllers.
20. The material collected in the vessel will show two separate phases, the upper one composed of excess oil and the lower one made up of microgels. Removing oil from the surface of microgels is sometimes challenging. The use of demulsifiers and/or organic solvents may be necessary.
21. Assessing the stability of microgels over time is essential to further address their use as DS. Choosing the correct concentration of the gelling agent to maintain the integrity of the microgels under the physiological conditions of the gastrointestinal tract is the key to the successful use of microgels as DS.
22. Some highly hydrophobic LPC may need to be dissolved in alcohol before being added to the oil to achieve complete solubilization. This dispersion must be subsequently heated to remove ethanol from the mixture. As heating this dispersion can cause oxidation of the BC, this process must be performed carefully at the lowest possible temperature (e.g., using low-pressure rotary evaporation).
23. Oil droplets dispersed in water are formed much more quickly than water droplets in oil because of the differences in viscosity of the continuous phases. Therefore, a different setting of flow rate values is expected compared to water-in-oil emulsions.

24. The material collected in the vessel will show two separate phases, the upper one composed of oil-in-water emulsion and the lower one the excess aqueous phase.
25. Characterize the emulsions in terms of size distribution and coefficient of variance.
26. Liposome size and polydispersity are affected by the concentration of phospholipids in the lipid phase. Preliminary bulk tests to determine the best formulation must be performed.
27. Liposomes are nanometric structures that cannot be observed using optical microscopy. Measuring the particle size distribution (e.g., by light scattering) is an efficient way to determine the best flow rate ratio to obtain liposomes with the desired features.
28. Flow-focusing of the organic phase causes a controlled diffusion of alcohol and water molecules between the streams, which favors the gradual replacement of alcohol by water molecules around the phospholipids. Consequently, the solubility of lipids in the system changes, triggering the self-organization of phospholipids into bilayer fragments and subsequently into spherical vesicles that trap the LPC between the hydrophobic tails of the phospholipids (Fig. 2c) [38–40].
29. The width of the lipid phase can be observed by optical microscopy and can give some insight into liposome formation (Fig. 3) [31]. A reduction in the aqueous/organic volumetric flow rate ratio promotes an increase in the width of the lipid stream, resulting in a central zone rich in an organic solvent. A higher concentration of solvent can stabilize the intermediate bilayers, allowing greater aggregation of phospholipids that generate vesicles with a wide size distribution. Conversely, by increasing the flow rate ratio, the organic stream will be focused into a narrower jet in which the organic solvent will be more diluted. This reduced concentration of organic solvent limits the formation of intermediate bilayers, resulting in smaller liposomes with a narrower size distribution [40–42].
30. Removal of ethanol is recommended to allow the application of liposomes in food products. This can be done by using a low-pressure rotary evaporator.

Acknowledgments

This research received financial support from the São Paulo Research Foundation (FAPESP) (Grants 2019/27354-3 and 2020/13217-1) and from the National Council of Research and Technological Development (CNPq) (Grant 307094/2021-9).

References

1. Whitesides GM (2006) The origins and the future of microfluidics. Nature 442:368–373. https://doi.org/10.1038/nature05058
2. Liu H, Singh RP, Zhang Z et al (2021) Microfluidic assembly: an innovative tool for the encapsulation, protection, and controlled release of nutraceuticals. J Agric Food Chem 69:2936–2949. https://doi.org/10.1021/acs.jafc.0c05395
3. Gkionis L, Aojula H, Harris LK, Tirella A (2021) Microfluidic-assisted fabrication of phosphatidylcholine-based liposomes for controlled drug delivery of chemotherapeutics. Int J Pharm 604. https://doi.org/10.1016/j.ijpharm.2021.120711
4. Fonseca LR, Santos TP, Czaikoski A, Cunha RL (2022) Microfluidics-based production of chitosan-gellan nanocomplexes encapsulating caffeine. Food Res Int 151. https://doi.org/10.1016/j.foodres.2021.110885
5. Caballero Aguilar LM, Duchi S, Onofrillo C et al (2021) Formation of alginate microspheres prepared by optimized microfluidics parameters for high encapsulation of bioactive molecules. J Colloid Interface Sci 587:240–251. https://doi.org/10.1016/j.jcis.2020.12.026
6. Sun BJ, Shum HC, Holtze C, Weitz DA (2010) Microfluidic melt emulsification for encapsulation and release of actives. ACS Appl Mater Interfaces 2:3411–3416. https://doi.org/10.1021/am100860b
7. Guimarães Sá Correia M, Briuglia ML, Niosi F, Lamprou DA (2017) Microfluidic manufacturing of phospholipid nanoparticles: stability, encapsulation efficacy, and drug release. Int J Pharm 516:91–99. https://doi.org/10.1016/j.ijpharm.2016.11.025
8. Wang W, Zhang MJ, Chu LY (2014) Microfluidic approach for encapsulation via double emulsions. Curr Opin Pharmacol 18:35–41. https://doi.org/10.1016/j.coph.2014.08.003
9. Santos TP, Costa ALR, Michelon M et al (2020) Development of a microfluidic route for the formation of gellan-based microgels incorporating jabuticaba (Myrciaria cauliflora) extract. J Food Eng 276. https://doi.org/10.1016/j.jfoodeng.2019.109884
10. Mark D, Haeberle S, Roth G et al (2010) Microfluidic lab-on-a-chip platforms: requirements, characteristics and applications. Chem Soc Rev 39:1153–1182. https://doi.org/10.1039/b820557b
11. Santos TP, Cejas CM, Cunha RL, Tabeling P (2021) Unraveling driving regimes for destabilizing concentrated emulsions within microchannels. Soft Matter 17:1821–1833. https://doi.org/10.1039/d0sm01674h
12. Zhang J, Xu W, Xu F et al (2021) Microfluidic droplet formation in co-flow devices fabricated by micro 3D printing. J Food Eng 290. https://doi.org/10.1016/j.jfoodeng.2020.110212
13. Jeong HH, Yelleswarapu VR, Yadavali S et al (2015) Kilo-scale droplet generation in three-dimensional monolithic elastomer device (3D MED). Lab Chip 15:4387–4392. https://doi.org/10.1039/c5lc01025j
14. Muluneh M, Issadore D (2013) Hybrid soft-lithography/laser machined microchips for the parallel generation of droplets. Lab Chip 13: 4750–4754. https://doi.org/10.1039/c3lc50979f
15. Kim SH, Kim JW, Kim DH et al (2013) Enhanced-throughput production of polymersomes using a parallelized capillary microfluidic device. Microfluid Nanofluid 14:509–514. https://doi.org/10.1007/s10404-012-1069-5
16. Feng Y, Lee Y (2019) Microfluidic assembly of food-grade delivery systems: toward functional delivery structure design. Trends Food Sci Technol 86:465–478. https://doi.org/10.1016/j.tifs.2019.02.054
17. ten Klooster S, Boerkamp VJP, Hennebelle M et al (2024) Unravelling the effect of droplet size on lipid oxidation in O/W emulsions by using microfluidics. Sci Rep 14. https://doi.org/10.1038/s41598-024-59170-9
18. Duncanson WJ, Lin T, Abate AR et al (2012) Microfluidic synthesis of advanced microparticles for encapsulation and controlled release. Lab Chip 12:2135–2145. https://doi.org/10.1039/c2lc21164e
19. Herranz-Blanco B, Ginestar E, Zhang H et al (2017) Microfluidics platform for glass capillaries and its application in droplet and nanoparticle fabrication. Int J Pharm 516:100–105. https://doi.org/10.1016/j.ijpharm.2016.11.024
20. Santos TP, Cejas CM, Cunha RL (2022) Microfluidics as a tool to assess and induce emulsion destabilization. Soft Matter 18:698–710. https://doi.org/10.1039/d1sm01588e
21. Deng B, Schroën K, de Ruiter J (2023) A microfluidic study of bubble formation and coalescence tuned by dynamic adsorption of

SDS and proteins. Food Hydrocoll 140. https://doi.org/10.1016/j.foodhyd.2023.108663
22. Sahin S, Schroën K (2015) Partitioned EDGE devices for high throughput production of monodisperse emulsion droplets with two distinct sizes. Lab Chip 15:2486–2495. https://doi.org/10.1039/c5lc00379b
23. Carugo D, Bottaro E, Owen J et al (2016) Liposome production by microfluidics: potential and limiting factors. Sci Rep 6. https://doi.org/10.1038/srep25876
24. Francisco CRL, Santos TP, Cunha RL (2023) Design of shear-based microfluidic channels for production and stability assessment of food emulsions. Curr Opin Food Sci 49. https://doi.org/10.1016/j.cofs.2022.100957
25. Mukhopadhyay R (2007) When PDMS isn't the best. Anal Chem 79:3248–3253
26. Adiraj Iyer M, Eddington DT (2019) Storing and releasing rhodamine as a model hydrophobic compound in polydimethylsiloxane microfluidic devices. Lab Chip 19:574–579. https://doi.org/10.1039/C9LC00039A
27. Yeh PY, Rossi NAA, Kizhakkedathu JN, Chiao M (2010) A silicone-based microfluidic chip grafted with carboxyl functionalized hyperbranched polyglycerols for selective protein capture. Microfluid Nanofluid 9:199–209. https://doi.org/10.1007/s10404-009-0535-1
28. Shakeri A, Khan S, Didar TF (2021) Conventional and emerging strategies for the fabrication and functionalization of PDMS-based microfluidic devices. Lab Chip 21:3053–3075. https://doi.org/10.1039/d1lc00288k
29. Zembyla M, Murray BS, Sarkar A (2020) Water-in-oil emulsions stabilized by surfactants, biopolymers and/or particles: a review. Trends Food Sci Technol 104:49–59. https://doi.org/10.1016/j.tifs.2020.07.028
30. Liu Z, Chai M, Chen X et al (2021) Emulsification in a microfluidic flow-focusing device: effect of the dispersed phase viscosity. Fuel 283. https://doi.org/10.1016/j.fuel.2020.119229
31. Michelon M, Oliveira DRB, de Figueiredo Furtado G et al (2017) High-throughput continuous production of liposomes using hydrodynamic flow-focusing microfluidic devices. Colloids Surf B Biointerfaces 156:349–357. https://doi.org/10.1016/j.colsurfb.2017.05.033
32. Duffy DC, McDonald JC, Schueller OJA, Whitesides GM (1998) Rapid prototyping of microfluidic systems in poly(dimethylsiloxane). Anal Chem 70:4974–4984. https://doi.org/10.1021/ac980656z
33. Nguyen HT, Marquis M, Anton M, Marze S (2019) Studying the real-time interplay between triglyceride digestion and lipophilic micronutrient bioaccessibility using droplet microfluidics. 1 lab on a chip method. Food Chem 275:523–529. https://doi.org/10.1016/j.foodchem.2018.09.096
34. Bodas D, Khan-Malek C (2006) Formation of more stable hydrophilic surfaces of PDMS by plasma and chemical treatments. Microelectron Eng 83:1277–1279. https://doi.org/10.1016/j.mee.2006.01.195
35. Chan LW, Lee HY, Heng PWS (2006) Mechanisms of external and internal gelation and their impact on the functions of alginate as a coat and delivery system. Carbohydr Polym 63:176–187. https://doi.org/10.1016/j.carbpol.2005.07.033
36. Chen M, Bolognesi G, Vladisavljević GT (2021) Crosslinking strategies for the microfluidic production of microgels. Molecules 26. https://doi.org/10.3390/molecules26123752
37. Kovalchuk NM, Roumpea E, Nowak E et al (2018) Effect of surfactant on emulsification in microchannels. Chem Eng Sci 176:139–152. https://doi.org/10.1016/j.ces.2017.10.026
38. Jahn A, Vreeland WN, Gaitan M, Locascio LE (2004) Controlled vesicle self-assembly in microfluidic channels with hydrodynamic focusing. J Am Chem Soc 126:2674–2675. https://doi.org/10.1021/ja0318030
39. Jahn A, Vreeland WN, Devoe DL et al (2007) Microfluidic directed formation of liposomes of controlled size. Langmuir 23:6289–6293. https://doi.org/10.1021/la070051a
40. Jahn A, Stavis SM, Hong JS et al (2010) Microfluidic mixing and the formation of nanoscale lipid vesicles. ACS Nano 4:2077–2087. https://doi.org/10.1021/nn901676x
41. Balbino TA, Aoki NT, Gasperini AAM et al (2013) Continuous flow production of cationic liposomes at high lipid concentration in microfluidic devices for gene delivery applications. Chem Eng J 226:423–433. https://doi.org/10.1016/j.cej.2013.04.053
42. Jahn A, Reiner JE, Vreeland WN et al (2008) Preparation of nanoparticles by continuous-flow microfluidics. J Nanopart Res 10:925–934

Chapter 12

Biosorption of Biomolecules from Vegetable Extracts Using Residual Yeast Cells

Fernanda Thaís Vieira Rubio, Charles Windson Isidoro Haminiuk, Giselle Maria Maciel, Júlia Cristina José, and Carmen Silvia Favaro-Trindade

Abstract

Residual yeasts from brewing or the production of ethanol represent a low-cost material and a source of proteins and β-glucans that can be enriched with bioactive compounds from vegetable extracts for food application. Biosorption is probably the most prominent phenomenon responsible for the incorporation of chemical compounds into yeasts. Thus, this protocol aimed to guide the reader along the key steps to the biosorption process using residual yeast biomass as sorbent and extracts as sorbate; it will help to follow specific steps, select important operational conditions, give an overview of the most critical factors (e.g., agitation speed, pH, and exposure to light and how to determine the correct equilibrium time), and provide tips on how to avoid some problems during the process (e.g., preventing the incorrect accounting of biosorbed compounds).

Key words *Saccharomyces cerevisiae*, *Saccharomyces pastorianus*, Biosorption kinetics, Isotherm models

1 Introduction

Biosorption is an adsorption process that uses biological materials as adsorbents (the reason for the prefix "bio"), such as microbial biomasses (bacteria, archaea, cyanobacteria, filamentous fungi, yeasts, and microalgae), seaweeds (macroalgae), and industrial and agricultural wastes [1]. Furthermore, it is a surface phenomenon that may involve the mass transfer of chemical substances from a liquid phase to a solid phase, which results in the removal or recovery of substances from aqueous media [2]. The first publication regarding biosorption was in 1951, by Ullrich and Smith [3], using the process in the treatment of domestic sewage. Thereafter, for several years, biosorption was primarily studied as a low-cost, high-efficiency, eco-friendly, and easy-design technique as a

Samantha Cristina de Pinho and Carmen S.F. Trindade (eds.), *Bioactives Encapsulation: Food Applications*, Methods and Protocols in Food Science, https://doi.org/10.1007/978-1-0716-4538-3_12,

solution for both the wastewater treatment and the reutilization of byproducts [4], once the generation of waste and byproducts from agroindustrial activities, along with wastewater pollution, remains a significant concern because of their environmental impact.

Over the last few years, yeasts have gained special attention beyond fermentation processes, and their cells have been used as important biosorbent material for the removal of pollutants such as heavy metals [5–8], pesticides [9], dyes [10–13], and antibiotics [14, 15]. The presence of functional groups in the cell wall of yeasts, such as hydroxyl, carboxyl, amino, sulfhydryl, ester, carbonyl, and phosphate, makes them proper for application as a biosorbent, considering the importance of the cell wall surface on the phenomenon of biosorption [16].

An advantage of using yeasts as biosorbents is that they represent an abundant byproduct from brewing and ethanol production and can be obtained in great amounts. In the brewing process, yeasts represent the second major byproduct, behind the solid residues generated from the processing of malt grain. Yeasts may be used in consecutive fermentation processes, depending on their viability, before being discarded [17]. Likewise, after the production of ethanol, yeasts are also part of the residues generated in the fermentation process [18]. The animal feed is still the main outlet for brewer's spent yeasts or residual yeasts from ethanol production, but they can still be explored, even when cells are not viable anymore.

In these last cases mentioned, yeasts were seen simply as a residue. Later, a relevant question came up: instead of using yeasts as "sponges" only to remove pollutants, why not use biosorption to enrich those yeasts with bioactive compounds to add value to a residue? This question made it clear that there was an underexplored potential in the food area, and several researchers started to enrich residual yeasts with bioactive compounds, such as phenolic compounds and anthocyanins [19–26]. Now, this byproduct represents a low-cost material obtained in abundant amounts that can encounter promising applications as a protein source, a source of β-glucans and mannoproteins for functional food applications, as a flavor-enhancing ingredient, and as an encapsulating and biosorption agent [27]. In addition, several works show that yeasts are able to protect the incorporated compounds during simulated gastrointestinal digestion [20, 21, 28–30], improving the compound bioaccessibility and highlighting the importance of yeasts as "delivery systems."

Although different methods have been used for the incorporation of compounds into yeasts, such as encapsulation by atomization [28, 31–34], emulsion [35–37], or vacuum infusion [38–40], all of them involve prior contact between the yeasts and the extract/emulsion, and it is probable that, in fact, the biosorption is the first and main process to occur. Thus, biosorption is possibly the foremost phenomenon responsible for the incorporation of compounds into yeast cells, and its study is of utmost importance. In this context, this protocol was written to help other authors carry out a complete biosorption study when incorporating chemical compounds into yeasts.

2 Material

Independently of whether the residual yeasts are obtained from brewing or the production of ethanol from sugarcane, the first step before using these yeasts is to wash them several times to remove beer or the sugarcane juice. This procedure is represented in Fig. 1. After receiving residual yeasts, it is important to store them under controlled temperature in a refrigerator at 4 °C. After that, yeasts will sediment, and it will be possible to remove the supernatant (beer or sugarcane juice). Distilled must be added to the sedimented yeasts in the proportion of 1:5 (v/v) of yeasts and water; they will be mixed manually until the suspension has a homogenous aspect, and a new sedimentation will occur at 4 °C.

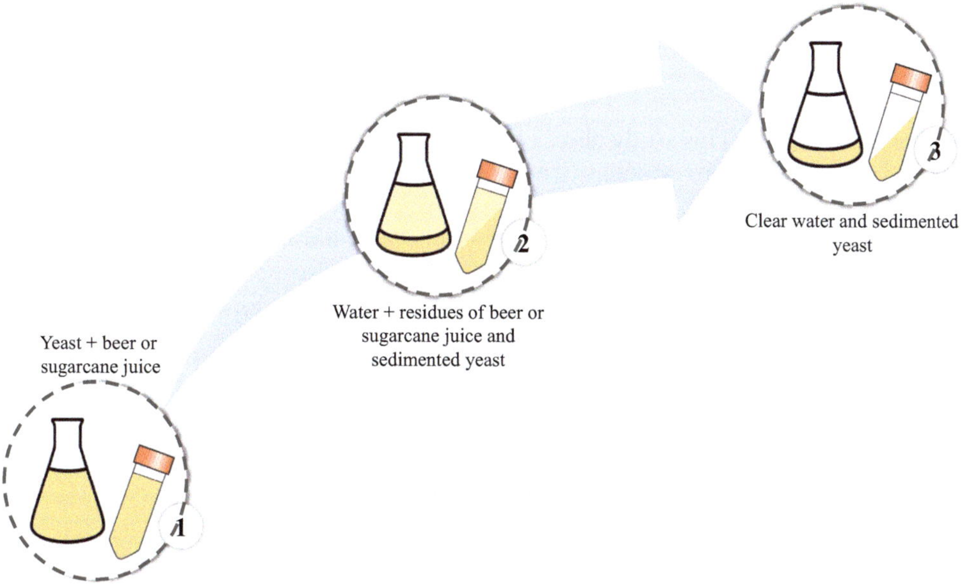

Fig. 1 Graphical representation of the washing and sedimentation steps of the yeasts where "*1*" represents how yeasts are received after their utilization for brewing or ethanol production. After their storage at 4 °C and waiting for their sedimentation, "*2*" represents the cells sedimented and separated from the residues of beer or sugarcane juice. Once these residues are removed, cells can be washed with distilled water, generally five times, until the water is clear. The last step is represented by "*3*"

The washing and sedimentation steps will be performed until the water is clear, without residues of beer or sugarcane juice. It is suggested to measure the reducing activity of sugar in this "washing water." Yeast washing can also be performed by centrifugation, firstly separating the material equally into Falcon conical tubes and operating a centrifuge at 4 °C and 4427×*g* for 5 min. Generally, five washings are required.

After washing, it is interesting to dehydrate the biomass to preserve the cells for a longer time and convenience. The drying of the washed cells can be performed both by spray-drying or freeze-drying. For freeze-drying, cells must be previously frozen. On the other side, spray-drying is faster, but it is necessary to add distilled water to the cells to decrease the viscosity and make it possible to pump the suspension through the atomizing nozzle. In this case, it is suggested to suspend the cells in water in the ratio of 1:5 (v/v) of wet yeast to distilled water. Furthermore, the suggested spray-drying conditions are an inlet temperature of 130 °C and an air-flow rate of 60 mL min^{-1}. During the process, the suspension must be kept under magnetic stirring to avoid yeast sedimentation during atomization.

After drying, cells may be stored under vacuum in polypropylene packages, at room temperature, or in a conventional refrigerator.

3 Methods

3.1 Biosorption Study

This study allows the biosorbent material (yeast biomass) to come into contact with the compound(s) of interest for their adsorption. This simple sorption dynamics test will establish the exposure time necessary for the system to equilibrate and determine the most preliminary information on the performance of any given sorption system. The sorption study can be done following the steps below:

1. Determine the concentration of the compound of interest in the extract before the biosorption study (e.g., by spectrophotometric or chromatographic methods) (*see* **Note 1**).
2. Select a maximum and a minimum time of contact between the biosorbent and the solution and some points in this range, for example, nine points: 360, 240, 120, 60, 30, 15, 10, 5, and 1 min. The number of samples in this experiment is equal to the number of points (times) selected.
3. Prepare the samples by blending a quantity of dry yeast with the solution containing the desired compounds. Weigh accurately the amount of biomass and record it. Ensure to regulate the pH of the system accordingly (*see* **Note 2**).

4. The experiment can occur in a thermostatic shaker, in a vortex mixer, or under magnetic or mechanical stirring (*see* **Note 3**). A suggested agitation speed is 150 rpm. The samples shall undergo stirring within a dark environment, particularly if the compounds of interest are light-sensitive. Adequate agitation is crucial to facilitate contact between the yeasts and the compounds within the solution.
5. In the process outlined, to remove all the samples simultaneously, the experiment should commence with the sample requiring the longest contact time with the yeast. For instance, the sample designated for 360 min should be prepared, placed in the equipment, and time should commence. Subsequently, after 120 min, the sample intended for 240 min should be introduced into the equipment, followed by consecutive additions in this manner.
6. Remove the samples and immediately centrifuge them to separate the yeast from the solution/extract (*see* **Note 4**). Collect the supernatant and use it to quantify the compounds. Each flask or tube represents one sample. It is not recommended to remove aliquots during the biosorption process.
7. The amount of compounds biosorbed shall be calculated using the following mass balance equation:

$$q = \frac{C_0 - C_f}{M} \times V \tag{1}$$

   where q is the sorption capacity (mg g^{-1}), C_0 and C_f are the concentrations of the compound in the initial extract and in the supernatant after biosorption, respectively. The volume (L) of the liquid phase is represented by V, and M is the amount of yeast used in the experiment (g).
8. Plot the results of the sorption capacity versus time and verify the last points of the experiment. The equilibrium is reached when the sorption capacity does not present any change. For example, if the sorption capacities at the points of 240 and 360 min are equal, the equilibrium time is 240 min (*see* **Note 5**)
9. Add blanks to the experiment. When working with extracts from natural sources, the compounds can be sensitive to several conditions, such as light, oxygen, extreme pH values, and high temperatures. Moreover, during a certain time of contact, the Erlenmeyer flasks can adsorb some compounds on their surface. Thus, if the compounds lost during the biosorption study are not discounted, they may be erroneously counted as being adsorbed on yeast. Therefore, it is important to (a) carry out an experiment without yeasts, only with the extracts, for the same times chosen previously for the biosorption study and

Fig. 2 Graphical representation of the biosorption study steps

(b) discount the lost compounds in each point from those quantified after biosorption.

10. It is recommended to carry out this experiment in triplicate.

Figure 2 shows a graphical representation of all these biosorption study steps.

3.2 Kinetic Study and Suggested Models

The kinetic study is carried out by varying the contact times between sorbent and sorbate, helping to understand the sorption speed and the factors that affect it, considering that the initial concentration of sorbent is reduced until dynamic equilibrium when the sites once available are filled and the amount of compound sorbed and desorbed is the same [41]. This study is also important for comparing experimental and theoretical values of sorption capacity and determining the applicability of kinetic models.

1. The results obtained in the biosorption study can be adjusted by the following models: pseudo-first-order model [42], pseudo-second-order model [43], intraparticle diffusion [44], Elovich model [45], and Blanchard model [46], respectively.

$$q_t = q_1 \times (1 - \exp(-k_1 \times t)) \quad (2)$$

$$q_t = \frac{k_2 \times q_2^2 \times t}{1 + k_2 \times q_2 \times t} \quad (3)$$

$$q_t = \left(K_d \times t^{0,5}\right) + C \quad (4)$$

$$q_t = \frac{1}{b} \times \ln(1 + a \times b \times t) \quad (5)$$

$$q_t = \frac{(k_4 \times q_4 \times t) + (a_4 \times q_4 - 1)}{k_4 \times t + a_4} \quad (6)$$

where q_t is the sorption capacity (mg g^{-1}) at time t (min); q_1 is the pseudo-first-order maximum adsorption capacity (mg g^{-1}); k_1 is the pseudo-first-order constant (min^{-1}); q_2 is the pseudo-second-order maximum adsorption capacity (mg g^{-1}); k_2 represents the pseudo-second-order constant (g mg^{-1} min^{-1}); k_d is the intraparticle diffusion coefficient (g mg^{-1} min$^{-0.5}$); C is the diffusion resistance constant (mg g^{-1}); a is the Elovich initial adsorption rate (mg g^{-1} min^{-1}); b is the Elovich's constant desorption rate (mg g^{-1}); q_4 is the Blanchard maximum adsorption capacity (mg g^{-1}); and k_4 and a_4 are Blanchard parameters.

2. The determination coefficient and the comparison between theoretical and experimental values of sorption capacities will indicate the model that provides the most accurate explanation for the sorption system. It is expected that the values of the determination coefficients are close to 1 and that the theoretical and experimental sorption capacities are as close as possible.

3.3 Sorption Isotherms

Once the equilibrium condition is reached in a system, the amount of a compound adsorbed per unit mass of biosorbent material can be described using sorption isotherms. Equilibrium relationships explain how the interaction between the biosorbent material and biosorbate occurs. Several isothermal models are applied to help understand the experimental process, comprehend how interactions occur, since each mathematical model considers different conditions: that biosorption can occur on homogeneous or heterogeneous surfaces, in monolayer or multilayer, and with chemical or physical interactions. Therefore, it becomes possible to describe the occurring phenomena, evaluating and optimizing the biosorption process [47]. The isotherm study can be carried out following the steps below:

1. Determine the initial compound concentration in the extract used for the isotherm study (also by spectrophotometric or chromatographic methods). This concentration will be your maximum concentration.
2. Prepare dilutions of this extract and adjust the pH. The recommendation is to prepare at least five different concentrations.
3. Weigh accurately the amount of biomass and record it.
4. Prepare the samples by adding the extracts to the biomass. The ratio of biosorbent and solution is the same as previously used in the biosorption study.

5. If the pH is an important parameter, make sure to control it during the experiment, taking care to not dilute samples by adding excessive volume of acid or base.
6. The experiment has to be carried out under the same conditions as the biosorption study (temperature, pH, agitation speed) and for the time required for the system to reach equilibrium. For example, if the equilibrium was reached in 240 min, your isotherm experiment will last this same time.
7. At the end of the experiment, separate the yeasts from the liquid by centrifugation.
8. Determine the compound concentration in the supernatants.
9. Adjust your data to mathematical models (*see* **Note 6**). Some of the models commonly used for this purpose and their nonlinear equations are presented in Table 1. The determination coefficient, chi-square (χ^2) and residual sum of squares (RSS) may be used to identify the best model to explain the biosorption process.
10. In the biosorption study, it is also important to add blanks to this experiment, which will contain only the extracts with different concentrations.

4 Notes

1. It is not suggested to use extracts prepared with absolute ethanol. Yeasts may dehydrate in ethanol, and the biosorption will not occur or not occur properly. Hydroalcoholic solutions are more appropriate.
2. The pH affects both the configuration of the active ion exchange sites as well as the ionic state of the sorbate in the solution. Thus, the pH control in the system during a biosorption process is important [60]. The pH affects the charge of the functional groups present in the cell wall and, therefore, affects the manner in which they will interact with the compounds in the solution.
3. Some alternative techniques can be used to improve or accelerate the biosorption process, such as ultrasound, electroporation, vacuum, and sorption under CO_2 supercritical conditions.
4. The yeasts obtained after their enrichment by biosorption can be dried, for example, by freeze- or spray-drying, and analyzed by techniques that help to prove their enrichment, such as scanning electron microscopy (SEM), confocal laser scanning microscopy (with or without staining the cells), transmission electron microscopy (TEM), X-ray diffraction spectroscopy

Table 1
Isotherm models are commonly used in biosorption studies

Model	Equation	References
Langmuir	$q_e = \frac{q_{max} K_L C_e}{1+K_L C_e},\ R_L = \frac{1}{1+K_L C_0}$	[48]
Freundlich	$q_e = K_F C_e^{1/n}$	[49]
Temkin	$q_e = \frac{RT}{b} \ln(a_t C_e)$	[50]
Dubinin-Radushkevich	$q_e = q_{max} \exp(-K_{dr}\varepsilon^2),\ \varepsilon = RT\left(1+\frac{1}{C_e}\right),\ E = \frac{1}{\sqrt{2K_{dr}}}$	[51]
Jovanovic Monolayer	$q_e = q_{max} \times (1 - \exp(-K_j \times C_e))$	[52]
Redlich-Peterson	$q_e = \frac{K_{RP} C_e}{1+\alpha_{RP} C_e^g}$	[53]
Hill	$q_e = \frac{q_{max} \times C_e^{nh}}{K_h + C_e^{nh}}$	[54]
Toth	$q_e = \frac{q_{max} \times C_e}{\left(b_{to} + C_e^{nt}\right)^{\frac{1}{nt}}}$	[55]
Sips	$q_e = \frac{q_{max} \times K_S \times C_e^S}{1+K_S \times C_e^S}$	[56]
Fritz-Schlünder	$q_e = \frac{A_{fs} \times C_e^{kfs}}{1+B_{fs}+C_e^{gfs}}$	[57]
Koble-Corrigan	$q_e = \frac{AC_e^{kc}}{\left(1+B_{KC} C_e^{kc}\right)}$	[58]
Radke-Prausnitz	$q_e = \frac{\left(K_{rp} q_{max} C_e\right)}{\left(1+K_{rp} C_e\right)^M}$	[59]

Here q_e is the amount of adsorbed solute per gram of biosorbent (mg g^{-1}) at equilibrium; q_{max} is the maximum adsorption capacity (mg g^{-1}); C_e is the concentration of adsorbate at equilibrium (mg L^{-1}); K_L is the Langmuir constant (L mg^{-1}); R_L is the Langmuir separation factor (dimensionless); n is the constant related to surface heterogeneity (dimensioless); K_F is the Freundlich adsorption capacity constant (L mg^{-1}); b is the angular coefficient from the linear relation of Temkin isotherm; a_t is the Temkin isotherm constant (L g^{-1}); ε is the Polanyi potential; K_{dr} is the constant associated with the adsorption energy (mol^2 $(KJ^2)^{-1}$); E (kJ mol^{-1}) is the adsorption free energy; K_j is the isotherm constant of Jovanovic (L mg^{-1}); K_{RP} is the Redlich-Peterson isothermal constant (L g^{-1}); α_{RP} is the Redlich-Peterson linear constant ($(mg\ L^{-1})^{-g}$); g is the Redlich-Peterson exponential constant, which lies in a range from 0 to 1; K_h is the Hill constant; nh is the cooperativity coefficient of the binding interaction; b_{to} and nt are Toth parameters; K_S (L mg^{-1}) is the equilibrium constant of Sips; s is the Sips exponential constant, which lies in the range from 0 to 1; A_{fs} is the adsorption constant; kfs, B_{fs}, and gfs are Fritz-Schlünder parameters; A, B_{KC}, and KC are the Koble-Corrigan parameters; K_{rp} is the Radke-Prausnitz equilibrium constant; M is the Radke-Prausnitz model exponent; R is the gas universal constant (8.314 J K^{-1} mol^{-1}); and T is the temperature (K).

(XRD), and Fourier-transform infrared spectroscopy (FTIR spectroscopy). Figure 3 shows an example of yeasts seen by confocal laser scanning microscopy.

5. If the biosorption study shows that the contact time between the yeast and the compound is not enough and the biosorption did not reach equilibrium, it is required to add more points to the biosorption study, increasing the time of the experiment. The equilibrium time depends on the affinity between the yeast

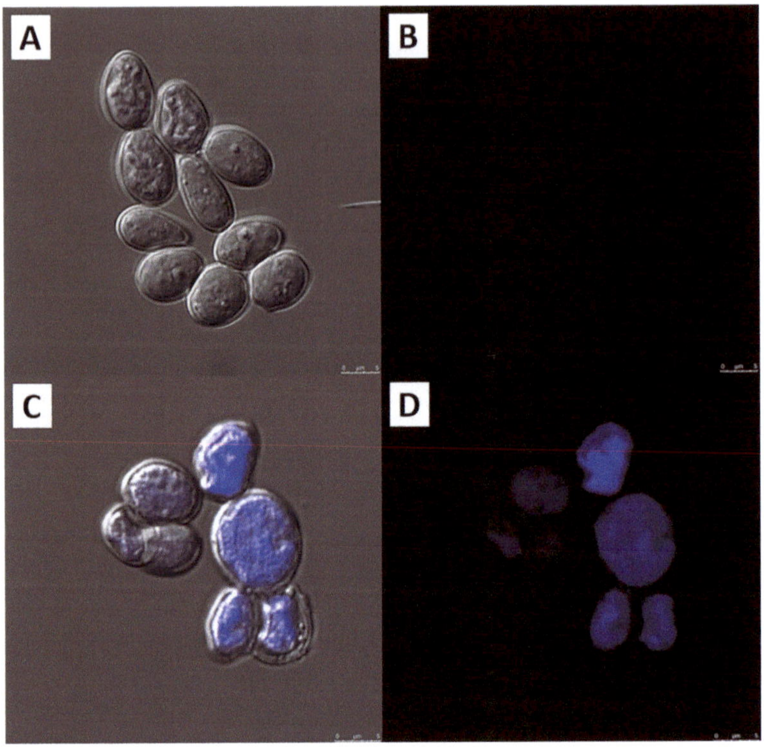

Fig. 3 Confocal laser scanning microscopy images of control and enriched yeasts. (**a**) and (**b**) represent control yeasts (non-enriched cells) in white light and without white light, respectively; (**c**) and (**d**) represent cells enriched with carotenoids from pumpkin peel extract, with and without white light, respectively. While control cells show no fluorescence under excitation, enriched cells show fluorescence, which indicates the probable incorporation of carotenoids inside the cells

and the biosorbent, the availability of the sites on the cell surface, and the compound concentration (to not saturate the sites on the yeast surface).

6. The suitability of mathematical isotherm models relies on the nature of the system. For instance, Langmuir is confined to monolayer adsorption, whereas the Freundlich model can accommodate multilayer adsorption. Additionally, both linear and nonlinear regression of these models can be employed to forecast the quantity of the adsorbed compound. However, not all isotherms adequately fit the experimental data and describe the biosorption process.

Acknowledgments

The authors would like to thank Conselho Nacional de Desenvolvimento Científico e Tecnológico (CNPq) and Fundação de Amparo à Pesquisa do Estado de São Paulo (FAPESP) for the post-doc fellowship granted to Fernanda Thaís Vieira Rubio (Processes 151120/2023-4 and 23/10387-1), FAPESP for the financial support (Process #21/12730-0), and Coordenação de Aperfeiçoamento de Pessoal de Nível Superior (CAPES) for the PhD scholarship granted to Julia Cristina José (Code 001).

References

1. Fomina M, Gadd GM (2014) Biosorption: current perspectives on concept, definition and application. Bioresour Technol [Internet] 160:3–14. Available from: https://doi.org/10.1016/j.biortech.2013.12.102
2. Chen X, Hossain F, Duan C, Lu J, Tsang YF, Islam S, et al. (2022) Isotherm models for adsorption of heavy metals from water – a review. [cited 2024 Apr 11]. Available from: www.elsevier.com/locate/chemospherehttps://doi.org/10.1016/j.chemosphere.2022.135545
3. Ullrich AH, Smith MW (1951) The biosorption process of sewage and waste treatment. Sewage Ind Waste 23(10):1248–1253
4. Wang J, Guo X, Tsang D, Ma LQ (2023) Adsorption kinetics and isotherm models of heavy metals by various adsorbents: an overview. Crit Rev Environ Sci Technol [Internet]. [cited 2024 Apr 11] 53(21):1837–1865. Available from: https://www.tandfonline.com/action/journalInformation?journalCode=best20
5. El-Bestawy EA, El-Batouti MM, Zabermawi NM, Zaghlol HM (2023) Removal of heavy metals, turbidity and coliform from contaminated raw drinking water using Saccharomyces cerevisiae, the Baker's yeast. [cited 2024 Mar 24]. Available from: https://doi.org/10.1016/j.scp.2023.101131
6. Chen C, Hu J, Wang J (2020) Biosorption of uranium by immobilized Saccharomyces cerevisiae. J Environ Radioact [Internet]. [cited 2024 Mar 24] 213:106158. Available from: https://doi.org/10.1016/j.jenvrad.2020.106158
7. Savastru E, Bulgariu D, Zamfir CI, Bulgariu L (2022) Application of Saccharomyces cerevisiae in the Biosorption of Co(II), Zn(II) and Cu(II) Ions from Aqueous Media. Water (Switzerland) 14(6):1–15
8. do Nascimento JM, de Oliveira JD, Rizzo ACL, Leite SGF (2019) Biosorption Cu (II) by the yeast Saccharomyces cerevisiae. Biotechnol Rep 21:e00315
9. Zhang M, Wen Y, Luo X, Wang X, Li J, Liu A et al (2021) Characterization, mechanism of cypermethrin biosorption by Saccharomyces cerevisiae strains YS81 and HP and removal of cypermethrin from apple and cucumber juices by inactive cells. J Hazard Mater 407:124350
10. Abbas Azeez R, Khaleel F, Al-Zuhairi I (2021) Biosorption of dye by immobilized yeast cells on the surface of magnetic nanoparticles. [cited 2024 Mar 24]. Available from: https://doi.org/10.1016/j.aej.2021.10.044
11. Soh EY, Shee Lim S, Wayne Chew K, Wei Phuang X, Ho VM, Chu KY, et al. (2022) Valorization of spent brewery yeast biosorbent with sonication-assisted adsorption for dye removal in wastewater treatment. Environ Res [Internet]. [cited 2024 Mar 24] 204:112385. Available from: https://doi.org/10.1016/j.envres.2021.112385
12. Zhang W, Zhang Y (2020). Development of ZnFe2O4 nanoparticle functionalized baker's yeast composite for effective removal of methylene blue via adsorption and photodegradation. [cited 2024 Apr 11]. Available from: https://doi.org/10.1016/j.jwpe.2020.101234
13. Mahmoud MS (2016) Decolorization of certain reactive dye from aqueous solution using Baker's Yeast (Saccharomyces cerevisiae) strain. HBRC J 12(1):88–98
14. Rusu L, Grigoraş CG, Simion AI, Suceveanu EM, Şuteu D, Harja M (2021) Application of saccharomyces cerevisiae/calcium alginate composite beads for cephalexin antibiotic biosorption from aqueous solutions. Materials (Basel) 14(16):4728

15. Samarghandi MR, Asgari G, Shokoohi R, Dargahi A, Arabkouhsar A (2019). Removing amoxicillin antibiotic from aqueous solutions by Saccharomyces cerevisiae bioadsorbent: kinetic, thermodynamic and isotherm studies. [cited 2024 Apr 11]. Available from: www.deswater.com
16. Jamir I, Ezung LY, Merry L, Tikendra L, Sanayaima Devi R, Nongdam P (2024) Heavy metals clean up: the application of fungi for biosorption. Geomicrobiol J [Internet]. [cited 2024 Mar 24] 41(3):201–212. Available from: https://doi.org/10.1080/01490451.2024.2307899
17. Ferreira IMPLVO, Pinho O, Vieira E, Tavarela JG (2010) Brewer's Saccharomyces yeast biomass: characteristics and potential applications. Trends Food Sci Technol [Internet] 21(2): 77–84. Available from: https://doi.org/10.1016/j.tifs.2009.10.008
18. Formann S, Hahn A, Janke L, Stinner W, Sträuber H, Logroño W, et al. (2020) Beyond sugar and ethanol production: value generation opportunities through sugarcane residues. Front Energy Res [Internet] [cited 2024 June 2] 8. Available from: www.frontiersin.org
19. Stafussa AP, Maciel GM, Da Silva Anthero AG, Da Silva MV, Zielinski AAF, Haminiuk CWI (2016) Biosorption of anthocyanins from grape pomace extracts by waste yeast: kinetic and isotherm studies. J Food Eng 169:53–60
20. Jilani H, Cilla A, Barberá R, Hamdi M (2016) Improved bioaccessibility and antioxidant capacity of olive leaf (Olea europaea L.) polyphenols through biosorption on Saccharomyces cerevisiae. Ind Crops Prod [Internet] 84: 131–138. Available from: https://doi.org/10.1016/j.indcrop.2016.02.002
21. Jilani H, Cilla A, Barberá R, Hamdi M (2015) Biosorption of green and black tea polyphenols into Saccharomyces cerevisiae improves their bioaccessibility. J Funct Foods 17:11–21
22. Rubio FTV, Maciel GM, da Silva MV, Corrêa VG, Peralta RM, Haminiuk CWI (2018) Enrichment of waste yeast with bioactive compounds from grape pomace as an innovative and emerging technology: Kinetics, isotherms and bioaccessibility. Innov Food Sci Emerg Technol [Internet] 45:18–28. Available from: http://linkinghub.elsevier.com/retrieve/pii/S1466856417304708
23. Rossetto R, Maciel GM, Bortolini DG, Ribeiro VR, Haminiuk CWI (2020) Acai pulp and seeds as emerging sources of phenolic compounds for enrichment of residual yeasts (Saccharomyces cerevisiae) through biosorption process. LWT - Food Sci Technol 128:109447
24. Ribeiro VR, Maciel GM, Fachi MM, Pontarolo R, de Fernandes IAA, Stafussa AP et al (2019) Improvement of phenolic compound bioaccessibility from yerba mate (Ilex paraguariensis) extracts after biosorption on Saccharomyces cerevisiae. Food Res Int [Internet] 126:108623. Available from: https://linkinghub.elsevier.com/retrieve/pii/S0963996919305010
25. de Oliveira ALMS, Maciel GM, Rossetto R, de Liz MV, Ribeiro VR, Haminiuk CWI (2019) Saccharomyces cerevisiae biosorbed with grape pomace flavonoids: adsorption studies and in vitro simulated gastrointestinal digestion. Int J Food Sci Technol [Internet] [cited 2021 July 19] 54(4):1413–1422. Available from: https://ifst.onlinelibrary.wiley.com/doi/full/10.1111/ijfs.14110
26. Ribeiro VR, Maciel GM, Fachi MM, Pontarolo R, de Andrade Arruda Fernandes I, Stafussa AP et al (2021) Biosorption of biocompounds from white and green tea in Saccharomyces cerevisiae waste: Study of the secondary metabolites by UPLC-QToF-MS and simulated in vitro gastrointestinal digestion. Food Biosci 41:101001
27. Jaeger A, Arendt EK, Zannini E, Sahin AW (2020) Brewer's spent yeast (BSY), an underutilized brewing by-product. Fermentation 6(4):123
28. Rubio FTV, Haminiuk CWI, Martelli-Tosi M, da Silva MP, Makimori GYF, Favaro-Trindade CS (2020) Utilization of grape pomaces and brewery waste Saccharomyces cerevisiae for the production of bio-based microencapsulated pigments. Food Res Int [Internet]. [cited 2020 July 1] 136:109470. Available from: https://linkinghub.elsevier.com/retrieve/pii/S0963996920304956
29. Rubio FTV, Haminiuk CWI, de Freitas Santos PD, Martelli-Tosi M, Thomazini M, de Carvalho Balieiro JC et al (2022) Investigation on brewer's spent yeast as a bio-vehicle for encapsulation of natural colorants from pumpkin (Cucurbita moschata) peels. Food Funct [Internet]. 13(19):10096–10109. Available from: https://xlink.rsc.org/?DOI=D2FO00759B
30. Ribeiro VR, Maciel GM, Fachi MM, Pontarolo R, de Fernandes IAA, Stafussa AP et al (2019) Improvement of phenolic compound bioaccessibility from yerba mate (Ilex paraguariensis) extracts after biosorption on Saccharomyces cerevisiae. Food Res Int 126: 108623
31. Sultana A, Miyamoto A, Lan Hy Q, Tanaka Y, Fushimi Y, Yoshii H (2017)

Microencapsulation of flavors by spray drying using Saccharomyces cerevisiae. J Food Eng 199:36–41

32. Sultana A, Tanaka Y, Fushimi Y, Yoshii H (2018) Stability and release behavior of encapsulated flavor from spray-dried Saccharomyces cerevisiae and maltodextrin powder. Food Res Int 106:809–816
33. Marson GV, Saturno RP, Comunian TA, Consoli L, da Machado MTC, Hubinger MD (2020) Maillard conjugates from spent brewer's yeast by-product as an innovative encapsulating material. Food Res Int 136: 109365
34. Rubio FTV, Haminiuk CWI, de Santos PDF, Martelli-Tosi M, Thomazini M, de Balieiro JCC et al (2022) Investigation on brewer's spent yeast as a bio-vehicle for encapsulation of natural colorants from pumpkin (Cucurbita moschata) peels. Food Funct [Internet] 13(19):10096–10109. Available from: http://xlink.rsc.org/?DOI=D2FO00759B
35. Beikzadeh S, Shojaee-Aliabadi S, Dadkhodazade E, Sheidaei Z, Abedi AS, Mirmoghtadaie L, et al (2020) View of comparison of properties of breads enriched with omega-3 oil encapsulated in β-glucan and saccharomyces cerevisiae yeast cells. Appl Food Biotechnol [Internet] [cited 2021 May 5] 7(1). Available from: https://journals.sbmu.ac.ir/afb/article/view/25969/Online
36. Fu J, Song L, Guan J, Sun C, Zhou D, Zhu B (2021) Encapsulation of Antarctic krill oil in yeast cell microcarriers: evaluation of oxidative stability and in vitro release. Food Chem 338: 128089
37. Dimopoulos G, Katsimichas A, Tsimogiannis D, Oreopoulou V, Taoukis P (2021) Cell permeabilization processes for improved encapsulation of oregano essential oil in yeast cells. J Food Eng 294:110408
38. Young S, Dea S, Nitin N (2017) Vacuum facilitated infusion of bioactives into yeast microcarriers: evaluation of a novel encapsulation approach. Food Res Int 100:100–112
39. Young S, Rai R, Nitin N (2020) Bioaccessibility of curcumin encapsulated in yeast cells and yeast cell wall particles. Food Chem [Internet] 309:125700. Available from: https://linkinghub.elsevier.com/retrieve/pii/S030881461931828X
40. Young S, Nitin N (2019) Thermal and oxidative stability of curcumin encapsulated in yeast microcarriers. Food Chem 275:1–7
41. Sen GS, Bhattacharyya KG (2011) Kinetics of adsorption of metal ions on inorganic materials: a review. Adv Colloid Interf Sci 162(1–2):39–58
42. Lagergren S (1898) Zur theorie der sogenannten adsorption geloster stoffe. Kungliga Svenska Vetenskapsakademiens Handlingar 24:1–39
43. Ho YS, McKay G (1999) Pseudo-second order model for sorption processes. Process Biochem 34:451–465
44. Weber WJ, Morris JC (1963) Kinetics of adsorption on carbon from solution. J Sanit Eng Div [Internet] 89(2):31–59. Available from: https://ascelibrary.org/doi/10.1061/JSEDAI.0000430
45. Low MJD (1960) Kinetics of Chemisorption of Gases on Solids. Chem Rev [Internet] 60(3):267–312. Available from: https://pubs.acs.org/doi/abs/10.1021/cr60205a003
46. Blanchard G, Maunaye M, Martin G (1984) Removal of heavy metals from waters by means of natural zeolites. Water Res 18(12): 1501–1507
47. Chen X, Yao M (2015) Modeling of experimental adsorption isotherm data. Information [Internet]. [cited 2024 May 13] 6(1):14–22. Available from: https://www.mdpi.com/2078-2489/6/1/14/htm
48. Langmuir I (1916) The constitution and fundamental properties of solids and liquids. Part I. Solids. J Am Chem Soc [Internet] 38(11): 2221–2295. Available from: http://pubs.acs.org/doi/abs/10.1021/ja02268a002
49. Freundlich HMF (1906) Over the adsorption in solution. J Phys Chem 57:385–471
50. Temkin MJ, Pyzhev V (1940) Kinetics of ammonia synthesis on promoted iron catalyst. Acta Physicochim USSR 12:217–222
51. Dubinin MM, Radushkevich LV (1947) The equation of the characteristic curve of activated charcoal. Proc Acad Sci Phys Chem Sect 55: 331–333
52. Jovanovic DS (1969) Physical adsorption of gases. I: isotherms for monolayer and multilayer adsorption. Colloid Polym Sci 235: 1203–1214
53. Redlich O, Peterson DL (1959) A useful adsorption isotherm. J Phys Chem [Internet] 63(6):1024–1024. Available from: https://pubs.acs.org/doi/abs/10.1021/j150576a611
54. Hill AV (1910) The possible effects of the aggregation of the molecules of haemoglobin on its dissociation curves. J Physiol 40:4–7
55. Tóth J (2000) Calculation of the BET-compatible surface area from any type I

isotherms measured above the critical temperature. J Colloid Interface Sci [Internet] 225(2): 378–383. Available from: https://linkinghub.elsevier.com/retrieve/pii/S0021979700967237

56. Sips R (1948) On the structure of a catalyst surface. J Chem Phys [Internet] 16(5): 490–495. Available from: https://pubs.aip.org/jcp/article/16/5/490/313266/On-the-Structure-of-a-Catalyst-Surface
57. Fritz W, Schlunder EU (1974) Dispersion of solutes in non-Newtonian laminar flow through a circular tube-eyring model fluid. Chem Eng Sci 29:1282–1286
58. Koble RA, Corrigan TE (1952) Adsorption isotherms for pure hydrocarbons. Ind Eng Chem [Internet] 44(2):383–387. Available from: https://pubs.acs.org/doi/abs/10.1021/ie50506a049
59. Radke CJ, Prausnitz JM (1972) Adsorption of organic solutes from dilute aqueous solution of activated carbon. Ind Eng Chem Fundam [Internet] 11(4):445–451. Available from: https://pubs.acs.org/doi/abs/10.1021/i160044a003
60. Volesky B (2007) Biosorption and me. Water Res 41(18):4017–4029

Chapter 13

Production of Yeast Biocapsules by Osmoporation

Fábio Gonçalves Macêdo de Medeiros,
Eduardo Wagner Vasconcelos de Andrade, Roberta Targino Hoskin, and Márcia Regina da Silva Pedrini

Abstract

Microencapsulation of bioactive compounds is a versatile and efficient strategy for promoting the stabilization and protection of labile, health-relevant molecules. Osmoporation is a promising technique for internalizing bioactive compounds utilizing yeast cells as a safe and efficient encapsulation system for multiple applications in the food and pharmaceutical industries. It is based on their natural osmoregulatory mechanisms that promote cell wall transient permeabilization as a response to sublethal osmotic perturbations. In this chapter, a detailed osmoporation protocol to produce yeast biocapsules is described, including the cultivation and harvest of food-grade yeast cells (*Saccharomyces cerevisiae*), preparation of osmomodulatory solutions, and single- and multistage osmoporation procedures.

Key words Osmoporation, Osmodulatory, Yeast cells, Permeabilization

1 Introduction

The growing consumer awareness and demand for healthier foods has boosted the industry's interest in the incorporation of naturally occurring, health-promoting bioactive compounds into food products [1]. However, food phytochemicals are generally unstable and susceptible to oxidation and degradation when exposed to food processing conditions and storage [2], which can compromise their bioavailability and potentially decrease their claimed health benefits [3]. Microencapsulation has long been regarded as a versatile strategy to stabilize and protect valuable, health-relevant bioactive compounds, and encapsulation systems using non-pathogenic, "generally recognized as safe" (GRAS) microorganisms have received increased attention as bio-based, sustainable, and versatile options [4, 5].

Samantha Cristina de Pinho and Carmen S.F. Trindade (eds.), *Bioactives Encapsulation: Food Applications*, Methods and Protocols in Food Science, https://doi.org/10.1007/978-1-0716-4538-3_13,

Osmoporation is an innovative and straightforward microencapsulation protocol based on cell membrane adaptive responses to osmotic perturbations in the medium that lead to temporary permeabilization of the cell wall structures [6, 7]. In this process, sublethal osmotic dehydration is induced by exposing the cells to a high osmotic pressure (π) medium, followed by shock rehydration at iso-osmotic conditions in the presence of the target bioactive compound to be encapsulated [6, 8]. During rehydration, the cell membrane undergoes transient permeabilization to allow cellular osmotic regulation, which results in the transfer of large volumes of water from the extracellular to the intracellular medium [6, 7, 9]. Both the intensity and the kinetics of the sublethal osmotic dehydration/rehydration cycles have been found to significantly affect the ability of the cells to promote osmotic regulation and, in turn, their viability [9], but osmotic shocks (rapid change in the medium osmotic pressure) have reportedly warranted the highest cell viability (>80%) and bioactive internalization efficiency [6, 10].

Contrary to time-consuming passive diffusion-based methods, in which the internalization of bioactive compounds into the cell structure is dependent on the bioactive concentration and natural interactions between the bioactive compound and the cell membrane to allow diffusion [4, 11], or to active endocytosis, a process that depends on energy expenditure and through which bioactive molecules would be located in vacuoles and possibly subjected to enzymatic degradation (Fig. 1a), osmoporation offers a rapid internalization of target phytochemicals by taking advantage of the significant mass transfer through the temporarily permeabilized membrane (Fig. 1b; [6, 10]). Moreover, Câmara Júnior et al. [10] demonstrated that during osmoporation, as a result of the active mass transfer through the permeabilized cell membrane, the bioactive compounds are delivered directly to the cell cytoplasm (Fig. 2), without the irreversible alterations to the cell membrane usually associated with other permeabilization protocols (e.g., cell plasmolysis, chemical hydrolysis, or electroporation; [12–14]).

Saccharomyces cerevisiae, also known as baker's yeast, is a low-cost, GRAS-status microorganism popularly used as an efficient bio-based microencapsulation system for large-scale production in the food industry [4, 8]. In addition to providing adequate stabilization and protection to the internalized bioactive compounds [15], *S. cerevisiae* cells also present robust osmoregulatory capacity [7, 9], making them ideal vehicles for producing osmoporation-based biocapsules. To date, osmoporation has been successfully applied for the encapsulation of hydrophilic high molecular weight fluorescent probe FITC (fluorescein isothiocyanate)-Dextran (20 kDa; [6]), low molecular weight hydrophobic flavonoid fisetin (286 Da; [10]) and low molecular weight hydrophobic polyphenol curcumin (368 Da [8]), through single and multistage processes [16] in *Saccharomyces cerevisiae*

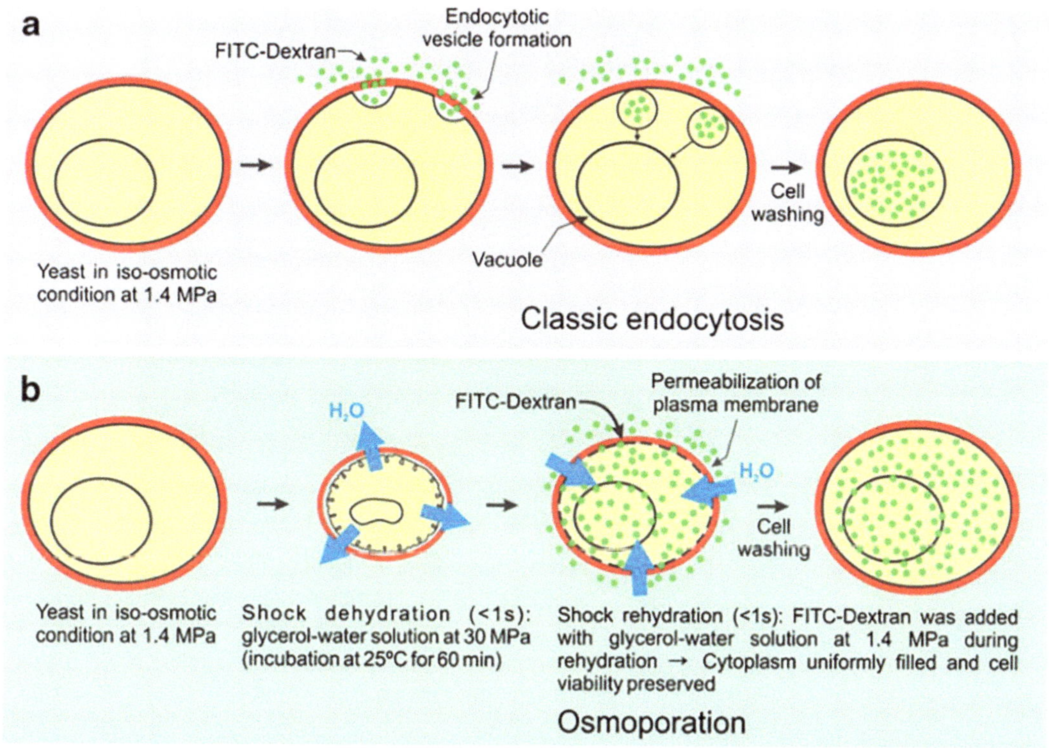

Fig. 1 Schematic representation of internalization steps of exogenous molecules in *Saccharomyces cerevisiae* cells during (**a**) classic iso-osmotic endocytosis and (**b**) osmoporation. (Reprinted with permission from Pedrini et al. [6]. Copyright Springer-Verlag Berlin Heidelberg 2013)

cells, and for the incorporation of fisetin into *Lactobacillus acidophilus* cells [5, 17]. In this chapter, we present a detailed protocol for the efficient encapsulation of bioactive compounds into yeast cells using osmoporation. Moreover, some relevant aspects of process optimization are presented and discussed.

2 Materials

2.1 Reagents

1. *Saccharomyces cerevisiae* viable culture (*see* **Note 1**) and bioactive compound of interest.
2. Glycerol (≥99%), ethanol (99°GL), distilled deionized (DDI) water.
3. Sodium chloride (NaCl, ≥98%), potassium chloride (KCl, ≥98%), disodium hydrogen phosphate (Na_2HPO_4, ≥98%), sodium dihydrogen phosphate (≥98%), potassium dihydrogen phosphate (KH_2PO_4, ≥98%).
4. Yeast extract, peptone, glucose (≥97%).
5. Sodium hydroxide (NaOH, ≥98%), hydrochloric acid (HCl, ≥36%).

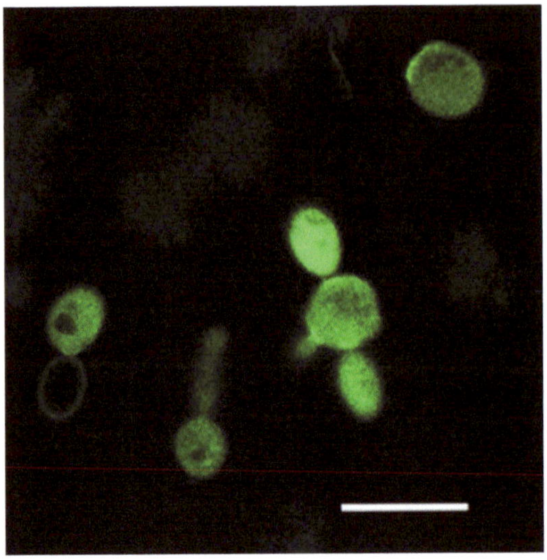

Fig. 2 Representative fluorescence confocal microscopy image of *Saccharomyces cerevisiae* cells submitted to osmoporation protocol (dehydration at 30 MPa for 60 min, followed by rehydration at 1.4 MPa for 60 min) for the internalization of fisetin at a final concentration of 2.0 mg/mL. (Reprinted with permission from Câmara Júnior et al [10]. Copyright Springer-Verlag Berlin Heidelberg 2016)

2.2 Media and Solutions

1. Bioactive compound solution: Prepare a solution of the bioactive compound of interest at a known concentration (*see* **Note 2**). Consider the solubility of the chosen bioactive molecule of the solution. DDI water, ethanol, or water–ethanol binary solutions are recommended.
2. Modified Malt-Wickerham medium (MW): In 800 mL of DDI water, add 3.0 g yeast extract, 3.0 g peptone, 10.0 g glucose, 51.0 g glycerol, and 1.5 g sodium dihydrogen phosphate (*see* **Note 3**). Dissolve all components and complete to a final volume of 1 L with DDI water. The modified MW medium should be sterilized prior to use for yeast cell cultivation.
3. Phosphate-buffered solution (1× PBS; pH 7.4): In 800 mL of DDI water, add 8.0 g NaCl, 0.2 g KCl, 1.44 g Na_2HPO_4, and 0.245 g KH_2PO_4. Dissolve all components and check that the buffer has reached pH 7.4. If necessary, adjust the pH using 0.1 M NaOH or 0.1 M HCl. Complete to a final volume of 1 L with DDI water and store at room temperature.

2.2.1 Glycerol–Water Binary Solutions

Glycerol ($C_3H_8O_3$, 1.26 g/mL) is used as an osmo-modulatory solute for the preparation of the hyperosmotic and iso-osmotic (1.4 MPa) solutions used on the shock dehydration and rehydration steps of the osmoporation protocol, respectively. The osmotic pressure (π) of the glycerol–water binary solutions is related to the

Table 1
Composition of glycerol–water binary solutions for different osmotic pressures

Osmotic pressure (π)	Water activity (a_w)	Glycerol (g/1000 g water)
1.4	0.99	51
5.0	0.96	181
10.0	0.93	360
15.0	0.90	531
20.0	0.86	700
30.0	0.80	1063
110.0	0.45	4005
166.0	0.30	6700

water activity (a_W) of the respective solution according to Eq. 1 [9]. The amount of osmo-modulatory solute (glycerol) can be calculated by the Norrish equation [18] in relation to the a_W of the respective solution, according to Eq. 2.

$$\pi\ (\text{Pa}) = \frac{-RT \ln a_W}{V_W} \quad (1)$$

$$a_W = (1 - X_s)e^{-KX_s^2} \quad (2)$$

where R is the universal gas constant (8.314 J/mol.K), T is the temperature (K), V_W is the molar volume of water (1.807×10^{-5} m^3/mol), X_s is the molar fraction of the glycerol in the glycerol–water binary mixture, and K is the Norrish coefficient of glycerol (1.16). Table 1 presents the amount of glycerol used to prepare glycerol–water binary mixtures with different useful osmotic pressures for osmoporation processes.

For the osmoporation of bioactive compounds into *S. cerevisiae* yeast cells, iso-osmotic (1.4 MPa) and hyperosmotic (30 MPa) glycerol–water binary solutions are used. Glycerol–water binary solutions should be sterilized prior to use for osmoporation protocols.

1. Iso-osmotic glycerol–water binary solution (1.4 MPa): Add 51 g of glycerol to 1000 g of water and mix thoroughly until complete dissolution.
2. Hyperosmotic glycerol–water binary solution (30 MPa): Add 1063 g of glycerol to 1000 g of water and mix thoroughly until complete dissolution.

2.3 Equipment

1. Laminar flow cabinet or other appropriate apparatus to ensure safe manipulation of yeast cultures.
2. Shaker incubator operating at 250 rpm, in the range of 25–30 °C.
3. Centrifuge for 50 mL tubes for cell harvest at 2200×*g* for 5 min at 4 °C (*see* **Note 4**).
4. Microcentrifuge for 2 mL tubes, for cell harvest at 5100×*g* for 10 min, at 4 °C (*see* **Note 4**).
5. Autoclave for media and solutions sterilization (121 °C, 20 min).
6. Ultra-freezer (−80 °C) and freeze dryer, for the evaluation of the encapsulation yield (*see* **Note 5**).

3 Methods

3.1 Cells Culture and Harvest Conditions

1. Transfer three previously isolated colonies of *S. cerevisiae* from modified MW medium plates to a conical flask with 100 mL of fresh MW medium (*see* **Note 6**).
2. Incubate at 25 °C, 250 rpm for 48 h in a shaker incubator.
3. Transfer 1 mL subcultures to 100 mL of fresh MW medium.
4. Incubate the subcultures at 25 °C, 250 rpm overnight (12–18 h), or until cell density reaches 10^8 cells/mL (*see* **Note 7**).
5. Take a subculture (40 mL) of cell suspension and centrifuge at 2200×*g*, for 5 min at 4 °C. Discard the supernatant.
6. Wash the cell pellet twice with 1.4 MPa iso-osmotic glycerol–water binary solution and resuspend the cells in 20 mL of the same solution.

3.2 Encapsulation via Osmoporation

The encapsulation of bioactive compounds into *S. cerevisiae* yeast cells is performed in two stages: the osmotic dehydration of the cells followed by the shock rehydration in the presence of the bioactive compound, as follows (Fig. 3).

1. Take 1.5 mL of cell suspension in iso-osmotic conditions (cell density 2×10^8 cells/mL) to a 2 mL microtube.
2. Centrifuge at 5100×*g* for 10 min at 4 °C and discard the supernatant.
3. Suspend the cell pellet in 1.5 mL of hyperosmotic 30 MPa glycerol–water binary solution. Mix vigorously to ensure complete resuspension of the cell pellet.
4. Incubate the cell suspension at 25 °C, 250 rpm for 60 min.

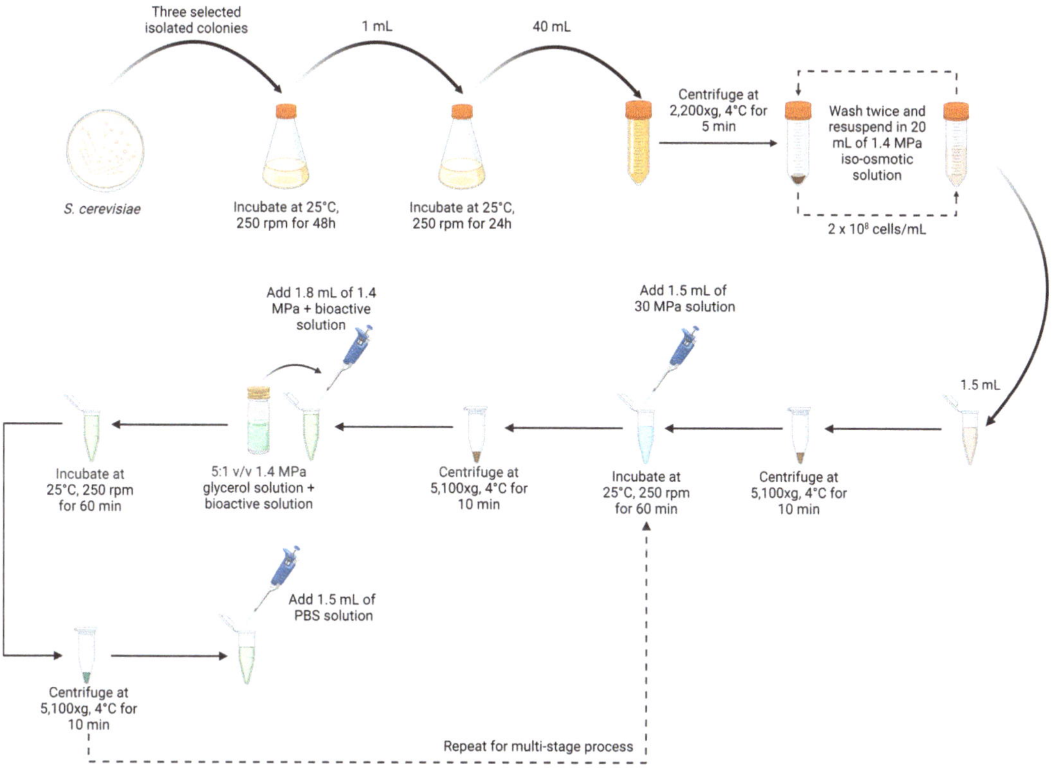

Fig. 3 Schematic representation of osmoporation protocol for the encapsulation of bioactive compounds into *Saccharomyces cerevisiae* cells

5. Centrifuge at 5100×*g* for 10 min at 4 °C and discard the supernatant.
6. Suspend the cell pellet in 1.5 mL of iso-osmotic 1.4 MPa glycerol–water binary solution, plus 0.3 mL of bioactive solution at a defined concentration (*see* **Note 8**). Mix vigorously to ensure complete resuspension of the cell pellet.
7. Incubate the resuspended cells in the presence of the bioactive compound at 25 °C, 250 rpm for 60 min (*see* **Note 9**).
8. Centrifuge at 5100×*g* for 10 min at 4 °C and discard the supernatant.
9. Suspend the cell pellet with 1.5 mL of PBS.

3.3 Evaluation of Encapsulation Process

The evaluation of the encapsulation of the process via osmoporation is carried out by direct extraction and quantification [10] and can be expressed in terms of encapsulation efficiency (EE, %) and encapsulation yield (EY, %; [8]).

3.3.1 Encapsulation Efficiency

For the quantification of encapsulation efficiency, perform the following steps immediately after the completion of the osmoporation protocol (**step 9**, Subheading 3.2).

1. Divide the recently prepared suspensions of yeast biocapsules into two groups of equal volume (or mass).
2. Transfer the first group to an ultrasonic bath (40 kHz, 25 °C) for 25 min for complete cell disruption and bioactive extraction. Dilute the samples to a final volume of 10 mL using ethanol and filter through a 0.45 μm nylon membrane.
3. The second group is not to be submitted to ultrasound cell disruption. Dilute the samples to a final volume of 10 mL using ethanol and filter through a 0.45 μm nylon membrane.
4. Bioactive concentration should be quantified with the appropriate method (*see* **Note 10**) Encapsulation efficiency can be calculated according to Eq. 3, where m_1 is the bioactive mass from the first group, submitted to ultrasound cell disruption, and m_2 is the bioactive mass from the second group, not submitted to this treatment.

$$\mathrm{EE}(\%) = \frac{m_1 - m_2}{m_1} \tag{3}$$

3.3.2 Encapsulation Yield

For the quantification of encapsulation yield, perform the following steps immediately after the completion of the osmoporation protocol (**step 9**, Subheading 3.2).

1. Centrifuge the cell suspension at 5100×*g* for 10 min at 4 °C and discard the supernatant.
2. Wash the cell pellet with cold PBS twice to ensure the removal of non-encapsulated bioactive compound.
3. Freeze to −80 °C and freeze-dry the cell suspensions (*see* **Note 5**).
4. Add 10 mg of the freeze-dried biocapsules to 10 mL of ethanol. Transfer the biocapsule suspension to an ultrasonic bath (40 kHz, 25 °C) for 25 min for complete cell disruption and bioactive extraction.
5. Centrifuge the cell suspension at 2200×*g* for 15 min at 4 °C.
6. Filter the supernatant through a 0.45 μm nylon membrane and proceed with the bioactive quantification with the appropriate method (*see* **Note 10**).
7. The encapsulation yield can be calculated according to Eq. 4, where m_B is the bioactive mass extracted from the dried biocapsules, and m_S is the mass of the dried biocapsules sample used for quantification.

$$\mathrm{EY}(\%) = \frac{m_B}{m_S} * 100 \tag{4}$$

3.4 Multistage Encapsulation

Multistage osmoporation has been demonstrated as an effective strategy for enhancing encapsulation performance. This process comprises sequential dehydration and rehydration stages, and it can decrease the amount of leftover, non-encapsulated bioactive, creating a rational way of taking full advantage of the encapsulation system (yeast cells; [16]). The following protocol describes the successful completion of a multistage (three stages) osmoporation protocol for the encapsulation of bioactive compounds into *S. cerevisiae* yeast cells (Fig. 3).

1. Proceed with cell cultivation, harvest, and one-step osmoporation as previously described in **steps 1–6**, Subheading 3.1, **steps 1–9**, Subheading 3.2.
2. Centrifuge at 5100×*g* for 10 min at 4 °C and collect the supernatant.
3. Suspend the cell pellet in 1.5 mL of hyperosmotic 30 MPa glycerol–water binary solution. Mix vigorously to ensure complete resuspension of the cell pellet.
4. Incubate the cell suspension at 25 °C, 250 rpm for 60 min.
5. Centrifuge at 5100×*g* for 10 min at 4 °C and collect the supernatant.
6. Suspend the cell pellet in the previously collected supernatant (1.4 MPa iso-osmotic solution plus bioactive solution) from the first rehydration step (*see* **Note 11**). Mix vigorously to ensure complete resuspension of the cell pellet.
7. Incubate the resuspended cells in the presence of the bioactive compound at 25 °C, 250 rpm for 60 min.
8. Centrifuge at 5100×*g* for 10 min at 4 °C and collect the supernatant.
9. Suspend the cell pellet in 1.5 mL of hyperosmotic 30 MPa glycerol–water binary solution. Mix vigorously to ensure complete resuspension of the cell pellet.
10. Incubate the cell suspension at 25 °C, 250 rpm for 60 min.
11. Centrifuge at 5100×*g* for 10 min at 4 °C and collect the supernatant.
12. Suspend the cell pellet in the previously collected supernatant (1.4 MPa iso-osmotic solution plus bioactive solution) from the second rehydration step. Mix vigorously to ensure complete resuspension of the cell pellet.
13. Incubate the resuspended cells in the presence of the bioactive compound at 25 °C, 250 rpm for 60 min.
14. Centrifuge at 5100×*g* for 10 min at 4 °C and discard the supernatant.
15. Suspend the cell pellet with 1.5 mL of PBS.

The quantification of encapsulation performance, namely, EE and EY, can be assessed by following **steps 1–4**, Subheading 3.3.1 and **1–7**, Subheading 3.3.2, respectively, immediately after the completion of **step 9**, Subheading 3.2 (first osmoporation stage), **step 3**, Subhading 3.4 (second osmoporation stage), or **step 10**, Subheading 3.4 (third osmoporation stage).

4 Notes

4.1 Osmoporation protocol

1. Viable *S. cerevisiae* cells are used for the osmoporation protocol. Both isolated strains (*S. cerevisiae* BY4742) and commercial food-grade *S. cerevisiae* have been investigated, and successful encapsulation performances using osmoporation have been reported.
2. To date, only a single compound encapsulation has been investigated using osmoporation. Complex bioactive-rich mixtures, such as polyphenol-rich extracts from fruits and other plant sources, have been used for the production of yeast biocapsules through other techniques and might also be used for osmoporation [4, 13, 19]. Key aspects regarding the preparation of the bioactive solution, namely, the bioactive concentration and the dilution solvent, have been found to significantly affect ($p < 0.05$) the efficiency of the encapsulation process via osmoporation [8, 10]. Bioactive concentrations from 0.32 to 5.0 mg/mL [5, 8, 10] and water–ethanol binary solutions [8] have been previously evaluated. The use of mild bioactive concentrations (around 2.0 mg/mL) and water–ethanol blends up to 60% v/v ethanol has been reported as the best condition for the encapsulation of low molecular weight hydrophobic molecules using osmoporation.
3. Glycerol (51 g/1000 g water) was introduced as a modification to MW medium to reach 1.4 MPa osmotic pressure, which has been recommended as the optimal osmotic pressure for the growth of *S. cerevisiae* [9] and matches the osmotic pressure of the iso-osmotic glycerol–water binary solution used for the osmoporation protocol.
4. The centrifugation of cell suspensions (2200×*g* for 5 min or 5100×*g* for 10 min) could be performed at 25 °C.
5. Alternatively, freezing of yeast biocapsules prior to freeze-drying could be performed at −20 °C or using liquid nitrogen.
6. Isolated *S. cerevisiae* colonies collected after 24 h of incubation (exponential phase) are recommended for use. If stock cultures are used, transfer 1 mL of a stock culture to 100 mL of fresh modified MW medium and incubate for 24 h at 25 °C, 250 rpm. Then, strike the culture in modified MW medium

plates (10 g/L agar), and incubate for 24 h at 25 °C for isolated colonies. If commercial food-grade *S. cerevisiae* is used, transfer 5 g of the dried culture to 100 mL of fresh modified MW medium and incubate for 24 h at 25 °C, 250 rpm. Transfer 1 mL subculture to 100 mL of fresh modified MW medium and incubate for an additional 24 h, at 25 °C, 250 rpm. Then, strike the culture in modified MW medium plates (10 g/L agar), and incubate for 24 h at 25 °C, for isolated colonies.

7. Cultivation conditions (temperature, time, agitation) might be adjusted according to the selected yeast strain (if other than *S. cerevisiae*). It is recommended to select appropriate cell cultivation conditions to ensure 10^8 cells/mL cell density and exponential growth phase at harvest.
8. At the rehydration step of the osmoporation protocol, the concentration of the bioactive solution in contact with the cells is diluted sixfold due to the addition of a 1.4 MPa iso-osmotic glycerol–water binary solution. Make sure to consider the dilution and calculate the concentration of the bioactive solution accordingly. Appropriate cell density can be confirmed by means of plating the solutions in a modified MW medium, direct counting in a Neubauer Hemocytometer (Neubauer chamber), or by flow cytometry, for example.
9. If the chosen bioactive is light sensitive, it is recommended to ensure light protection during rehydration, for example, by wrapping tubes with aluminum foil.
10. Previous reports on single compound encapsulation using osmoporation have reported the use of spectrophotometry as a means of quantifying the concentration of the target bioactive compounds. Fisetin [10] and curcumin [8] were quantified using external calibration curves with absorbance read at 360 and 425 nm, respectively.
11. For sequential, multistage osmoporation, the supernatant from one osmoporation step containing the leftover, non-encapsulated bioactive compound diluted in the 1.4 MPa iso-osmotic glycerol–water binary solution is used as a rehydration solution for the following osmoporation step. Measurements should be taken to maintain the integrity of the bioactive compound in the supernatant during the dehydration step. For example, if the bioactive compound is photosensitive, the supernatant solution should be stored away from the light until use.
12. Osmotic perturbations have been reported to affect the viability of yeast cells. It depends on the level and kinetics of the osmotic changes and the structural modifications in the yeast cell membrane resulting from osmoregulatory mechanisms [9, 20]. It has been extensively demonstrated that cell viability

remains above 80% when yeast cells are submitted to 30 MPa osmotic shock for 60 min [6, 10]. Dupont et al. [9] highlighted that an osmotic event at 30 MPa can arrest cell growth, but the integrity of the cell membrane is preserved, even with the maintenance of the osmotic pressure (30 MPa) up to 60 min. However, sequential osmoporation cycles can negatively affect the viability of yeast cells. Medeiros et al. [16] showed that three sequential osmoporation stages (dehydration at 30 MPa followed by rehydration at 1.4 MPa, repeated three times) decreased cell viability to around 15%, which might negatively affect the effectiveness of further sequential osmoporation stages for the encapsulation of bioactive compounds.

13. The performance of osmoporation for the encapsulation of bioactive compounds (EE, EY) is affected by different process conditions, including bioactive concentration, cell density, and osmotic pressure of the dehydration step and incubation temperature. The available data on the optimization of bioactive encapsulation via osmoporation is related to the encapsulation of low molecular weight hydrophobic molecules (fisetin and curcumin; [8, 10]).
14. 14. Increasing cell density was found to positively affect the encapsulation performance of osmoporation. In vivo encapsulating systems use pre-defined structures, such as yeast cells, as encapsulating agents constrained by the space available in such preformed structures [7]. In this sense, increasing cell density has been directly associated with an increase in the space available for bioactive encapsulation and, therefore, has been shown to improve encapsulation performance [8, 21]. On the other hand, high concentrations of bioactive compounds have been found deleterious to the encapsulation performance of in vivo encapsulation systems, such as yeast cells [21]. This negative trend might be related to possible irreversible changes or disruption in the yeast cell envelope when in contact with the high osmolarity of highly concentrated bioactive solutions [20], or to possible toxicity-related structural changes in the yeast cell membrane, e.g., pore formation, emerging from similarly high bioactive concentrations [22]. Mild concentrations of both fisetin and curcumin, around 2.0 mg/mL, were found to be the optimal range for increasing encapsulation performance by osmoporation [8, 10]. Furthermore, the use of water–ethanol blends as bioactive solvent carriers has been proven to reduce the bioactive molecule aggregates, promoting proper solubilization in the encapsulation medium, and enlarging the pores' size on the yeast structures' surface for better molecule loading [23, 24]. Overall, optimal encapsulation in yeast biocapsules was obtained with ethanol contents between 35% and 60% [8, 21, 25].

15. The osmotic pressure used for cell dehydration and the incubation temperature during the dehydration/rehydration steps influence the encapsulation performance of osmoporation because both these parameters affect the yeast cell membrane structure [10]. During osmotic dehydration, part of the intracellular volume flows to the outside of the cell [26] as a response to the osmotic perturbation. On the contrary, during rehydration, the cell membrane is temporarily permeabilized to allow an inflow of extracellular water [27] that carries extracellular components and encapsulates them within the yeast cell structure [6].
16. The osmotic pressure of dehydration has a direct influence on the degree of reorganization of the yeast cell membrane as a reaction to this osmotic event [9]. When high osmotic pressures (>30 MPa) are applied during dehydration, the overall integrity of the cell membrane tends to be compromised, as cell permeabilization might be too severe and arrest the membrane's reorganization ability, leading to leaking of the cellular content [7, 9]. Similarly, incubation temperature above 25 °C might lead to increased membrane permeabilization as a result of enhanced proton flow and activity of H^+-ATPase, which can also lead to severe and irreversible membrane permeabilization [10]. The compromise of the yeast cell membrane integrity negatively affects the efficiency of the encapsulation process, as the yeast cell no longer has the capacity to hold in the intracellular contents and, therefore, cannot act as an efficient encapsulation system.

Osmoporation is a simple, versatile, and straightforward procedure for the internalization of bioactive compounds into yeast cells. Overall, mild bioactive concentrations, high cell density, and sublethal osmotic pressures and temperatures were reported to enhance the encapsulation performance of osmoporation. Process optimization trials, including the use of response surface methodology, have been demonstrated to be useful techniques to maximize encapsulation efficiency and yield and take full advantage of yeast cells as efficient food-grade encapsulation systems for osmoporation.

References

1. Malek L, Umberger WJ (2023) Protein source matters: understanding consumer segments with distinct preferences for alternative proteins. Future Foods 7. https://doi.org/10.1016/j.fufo.2023.100220
2. Hoskin RT, Grace MH, Xiong J, Lila MA (2023) Spray-drying microencapsulation of blackcurrant and cocoa polyphenols using underexplored plant-based protein sources. J Food Sci 88:2665–2678. https://doi.org/10.1111/1750-3841.16590
3. Lila MA, Hoskin RT, Grace MH et al (2022) Boosting the bioaccessibility of dietary bioactives by delivery as protein–polyphenol aggregate particles. J Agric Food Chem 70:13017–

13026. https://doi.org/10.1021/acs.jafc.2c00398

4. Rubio FTV, Haminiuk CWI, dos Santos MM et al (2021) Development of natural pigments microencapsulated in waste yeast *Saccharomyces cerevisiae* using spray drying technology and their application in yogurt. Food Funct 12:8946–8959. https://doi.org/10.1039/D1FO00708D
5. de Andrade EWV, Dupont S, Beney L et al (2022) Osmoporation is a versatile technique to encapsulate fisetin using the probiotic bacteria lactobacillus acidophilus. Appl Microbiol Biotechnol 106:1031–1044. https://doi.org/10.1007/s00253-021-11735-8
6. da Silva Pedrini MR, Dupont S, De Anchieta Câmara A et al (2014) Osmoporation: a simple way to internalize hydrophilic molecules into yeast. Appl Microbiol Biotechnol 98:1271–1280. https://doi.org/10.1007/s00253-013-5386-8
7. Dupont S, Rapoport A, Gervais P, Beney L (2014) Survival kit of Saccharomyces cerevisiae for anhydrobiosis. Appl Microbiol Biotechnol 98:8821–8834
8. de Medeiros FGM, Dupont S, Beney L et al (2019) Efficient stabilisation of curcumin microencapsulated into yeast cells via osmoporation. Appl Microbiol Biotechnol 103:9659–9672. https://doi.org/10.1007/s00253-019-10196-4
9. Dupont S, Beney L, Ritt JF et al (2010) Lateral reorganization of plasma membrane is involved in the yeast resistance to severe dehydration. Biochim Biophys Acta Biomembr 1798:975–985. https://doi.org/10.1016/j.bbamem.2010.01.015
10. Câmara Júnior AA, Dupont S, Beney L et al (2016) Fisetin yeast-based bio-capsules via osmoporation: effects of process variables on the encapsulation efficiency and internalized fisetin content. Appl Microbiol Biotechnol 100:5547–5558. https://doi.org/10.1007/s00253-016-7425-8
11. Dadkhodazade E, Mohammadi A, Shojaee-Aliabadi S et al (2018) Yeast cell microcapsules as a novel carrier for cholecalciferol encapsulation: development, characterization and release properties. Food Biophys 13:404–411. https://doi.org/10.1007/s11483-018-9546-3
12. Dimopoulos G, Katsimichas A, Tsimogiannis D et al (2021) Cell permeabilization processes for improved encapsulation of oregano essential oil in yeast cells. J Food Eng 294:110408. https://doi.org/10.1016/j.jfoodeng.2020.110408
13. Rubio FTV, Maciel GM, da Silva MV et al (2018) Enrichment of waste yeast with bioactive compounds from grape pomace as an innovative and emerging technology: kinetics, isotherms and bioaccessibility. Innovative Food Sci Emerg Technol 45:18–28. https://doi.org/10.1016/j.ifset.2017.09.004
14. Kurek MA, Majek M, Onopiuk A et al (2023) Encapsulation of anthocyanins from chokeberry (Aronia melanocarpa) with plazmolyzed yeast cells of different species. Food Bioprod Process 137:84–92. https://doi.org/10.1016/j.fbp.2022.11.001
15. Paramera EI, Karathanos VT, Konteles SJ (2023) Yeast cells and yeast-based materials for microencapsulation. In: Microencapsulation in the food industry. Elsevier, Londres, pp 343–365
16. Medeiros FGM, Correia RTP, Dupont S et al (2018) Curcumin and fisetin internalization into Saccharomyces cerevisiae cells via osmoporation: impact of multiple osmotic treatments on the process efficiency. Lett Appl Microbiol 67:363–369. https://doi.org/10.1111/lam.13045
17. de Andrade EWV, Dupont S, Beney L et al (2022) Techno-functionality of fisetin-enriched yoghurt fermented with lactobacillus acidophilus bio-capsules produced via osmoporation. Syst Microbiol Biomanuf 2:743–749. https://doi.org/10.1007/s43393-022-00100-z
18. Norrish RS (1966) An equation for the activity coefficients and equilibrium relative humidities of water in confectionery syrups. Int J Food Sci Technol 1:25–39. https://doi.org/10.1111/j.1365-2621.1966.tb01027.x
19. de Andrade EWV, Hoskin RT, Dupont S et al (2024) Sonoprocessing coupled to spray drying as a novel strategy to encapsulate bioactive compounds from acerola pomace extract into Saccharomyces cerevisiae cells. Syst Microbiol Biomanuf. https://doi.org/10.1007/s43393-024-00248-w
20. Simonin H, Beney L, Gervais P (2007) Sequence of occurring damages in yeast plasma membrane during dehydration and rehydration: mechanisms of cell death. Biochim Biophys Acta Biomembr 1768:1600–1610. https://doi.org/10.1016/j.bbamem.2007.03.017
21. Paramera EI, Konteles SJ, Karathanos VT (2011) Stability and release properties of curcumin encapsulated in Saccharomyces cerevisiae, β-cyclodextrin and modified starch. Food Chem 125:913–922. https://doi.org/10.1016/j.foodchem.2010.09.071

22. de Souza Pereira R, Geibel J (1999) Direct observation of oxidative stress on the cell wall of Saccharomyces cerevisiae strains with atomic force microscopy. Mol Cell Biochem 201:17–24. https://doi.org/10.1023/A:1007007704657
23. Pham-Hoang B-N, Phan-Thi H, Waché Y (2015) Can biological structures be natural and sustainable capsules? Front Chem 3. https://doi.org/10.3389/fchem.2015.00036
24. Karaman K (2021) Fabrication of gallic acid loaded yeast (Saccharomyces cerevisiae) microcapsules: effect of plasmolysis treatment and solvent type on bioactivity and release kinetics. LWT 148:111640. https://doi.org/10.1016/j.lwt.2021.111640
25. Young S, Dea S, Nitin N (2017) Vacuum facilitated infusion of bioactives into yeast microcarriers: evaluation of a novel encapsulation approach. Food Res Int 100:100–112. https://doi.org/10.1016/j.foodres.2017.07.067
26. Gervais P, Beney L (2001) Osmotic mass transfer in the yeast Saccharomyces cerevisiae. Cell Mol Biol (Noisy-le-Grand) 47:831–839
27. Schaber J, Adrover MÀ, Eriksson E et al (2010) Biophysical properties of Saccharomyces cerevisiae and their relationship with HOG pathway activation. Eur Biophys J 39:1547–1556. https://doi.org/10.1007/s00249-010-0612-0

Index

Samantha Cristina de Pinho and Carmen S.F. Trindade (eds.), *Bioactives Encapsulation: Food Applications*, Methods and Protocols in Food Science, https://doi.org/10.1007/978-1-0716-4538-3,